T0219861

Horticultural Therapy Methods

Connecting People and Plants in Health Care, Human Services, and Therapeutic Programs

SECOND EDITION

Horticultural Therapy Methods

Connecting People and Plants in Health Care, Human Services, and Therapeutic Programs

SECOND EDITION

EDITORS

Rebecca L. Haller • Christine L. Capra

CRC Press
Taylor & Francis Group
Boca Raton London New York

CRC Press is an imprint of the
Taylor & Francis Group, an **informa** business

CRC Press
Taylor & Francis Group
6000 Broken Sound Parkway NW, Suite 300
Boca Raton, FL 33487-2742

© 2017 by Taylor & Francis Group, LLC
CRC Press is an imprint of Taylor & Francis Group, an Informa business

No claim to original U.S. Government works

Printed on acid-free paper
Version Date: 20160531

International Standard Book Number-13: 978-1-138-73117-2 (Paperback) 978-1-4987-3699-2 (Hardback)

This book contains information obtained from authentic and highly regarded sources. Reasonable efforts have been made to publish reliable data and information, but the author and publisher cannot assume responsibility for the validity of all materials or the consequences of their use. The authors and publishers have attempted to trace the copyright holders of all material reproduced in this publication and apologize to copyright holders if permission to publish in this form has not been obtained. If any copyright material has not been acknowledged please write and let us know so we may rectify in any future reprint.

Except as permitted under U.S. Copyright Law, no part of this book may be reprinted, reproduced, transmitted, or utilized in any form by any electronic, mechanical, or other means, now known or hereafter invented, including photocopying, microfilming, and recording, or in any information storage or retrieval system, without written permission from the publishers.

For permission to photocopy or use material electronically from this work, please access www.copyright. com (http://www.copyright.com/) or contact the Copyright Clearance Center, Inc. (CCC), 222 Rosewood Drive, Danvers, MA 01923, 978-750-8400. CCC is a not-for-profit organization that provides licenses and registration for a variety of users. For organizations that have been granted a photocopy license by the CCC, a separate system of payment has been arranged.

Trademark Notice: Product or corporate names may be trademarks or registered trademarks, and are used only for identification and explanation without intent to infringe.

Library of Congress Cataloging-in-Publication Data

Names: Haller, Rebecca L., editor.
Title: Horticultural therapy methods : connecting people and plants in health
care, human services, and therapeutic programs / editors: Rebecca L.
Haller & Christine L. Capra.
Description: Second edition. | Boca Raton : Taylor & Francis, 2017. |
Includes bibliographical references and index.
Identifiers: LCCN 2016023043| ISBN 9781138731172 (pbk. : alk. paper) | ISBN
9781498736992 (hardback : alk. paper) | ISBN 9781498732888 (e-book)
Subjects: LCSH: Gardening--Therapeutic use.
Classification: LCC RM735.7.G37 H675 2017 | DDC 615.8/515--dc23
LC record available at https://lccn.loc.gov/2016023043

Visit the Taylor & Francis Web site at
http://www.taylorandfrancis.com

and the CRC Press Web site at
http://www.crcpress.com

Dedication

*To Jack Kramer for showing me the power of the garden
and Kateri and Hannah who continue his legacy and
participate in the therapeutic healing that nature brings.*

Christine L. Capra

*To my parents, Deloris and Edward Haller, and
my grandparents, William and Sophia Thowe, who
introduced me to the wonders of the garden.*

Rebecca L. Haller

Contents

Foreword

That horticultural therapy is maturing into a valued member of the international health-care community is evident from the content and quality of this book. Having had the opportunity over the past 50 years to observe and at times participate in the gradual evolution of horticultural therapy—from an activity conducted exclusively by volunteer gardeners to a recognized therapeutic modality that includes trained, registered professionals—it is gratifying to see the emergence of such a clear, concise book that meets the needs of both professionals and volunteers. This book will be useful to educators in the health-care arenas and the horticulture field as it provides an excellent framework for communicating the basics of horticultural therapy. At the same time, it leads readers to other resources for information and challenges them to meet the needs that exist for future growth of the field.

The authors are experienced practitioners and educators. As such, they have been able to identify and explore key issues that will help individuals coming into this field—from health care or horticulture, as professionals or volunteers—to understand the theory, application, and impact of their work. This book is a valuable contribution to the growth of horticultural therapy, and I would like to commend the authors for the dedication it required to gain the skills and knowledge in horticultural therapy and to share them through their work.

Paula Diane Relf
Professor Emeritus, Virginia Tech University
Former Chair, People Plant Council

Preface

This book is a revision of the first edition published in 2006. Both editions were written to increase the value of using horticulture for therapy and human development by describing processes and techniques for practice. Chapters have been revised and expanded to reflect current terminology and practice in the field of horticultural therapy, including vocational, therapeutic, and wellness programs. A new chapter on treatment sessions guides the therapist in essential skills in this area. Appendices have been significantly revised, most notably with the addition of extensive horticultural therapy treatment strategies. The inclusion of photos enhances the experience of reading the book and shows the variety of horticultural therapy sites and types of practice.

Based on many years of experience and teaching, the editors hope to encourage competent use of the processes and techniques which are described in this book and which are widely accepted by allied professions. The manual is geared toward repeated use as a reference and to help students, educators, and those conducting horticultural therapy to provide effective and respected programs. It is also aimed at a variety of health-care and human service professionals who use horticultural therapy as well as at community horticulture program leaders and serious nonprofessionals such as volunteers and master gardeners.

For simplicity, the term *client* is used throughout the text to indicate the person served by the horticultural therapy program. It is intended as a catchall term to refer to the consumer, inmate, resident, patient, trainee, student, or anyone who participates in horticultural therapy (and at whom the program is directed). The reader is reminded that the client is first and foremost a unique person rather than a label. The use of "HT" as an abbreviation for "horticultural therapy" or "horticultural therapist" may be found in some areas of the text for brevity. The terms *therapeutic horticulture* and *community horticulture* are used to describe horticulture programs that exist to support human or community development but don't necessarily utilize treatment planning or goals. Leaders of these types of programs may not necessarily utilize all of the methods described in this book, yet will benefit from planning, leading, and documenting outcomes based on the procedures outlined.

The text is organized into six chapters, with supplemental material included in appendices. Chapter 1 sets the framework for discussion of the techniques in later sections. It also provides background on the shaping of the profession and how horticultural therapy is defined. Chapter 2 outlines the treatment process used in horticultural therapy programs. Chapter 3 instructs the reader how to go about activity planning, and choosing and scheduling horticultural tasks. Chapter 4 covers information about working with program participants and some techniques that are important for therapists, trainers, and program facilitators. The addition of Chapter 5 takes the reader through the process of planning a horticultural therapy treatment session. Last, Chapter 6 provides reasoning and directions for the documentation of treatment processes and outcomes. The appendices include forms and examples of documents used in the treatment process as well as ideas for writing objectives and choosing relevant activities and treatment strategies. It is hoped that readers will use these examples as aids to create appropriate sessions and documents for the setting in which they practice.

At the end of each chapter, references are provided for the reader to further explore the topics covered. The manual focuses on skills involving treatment planning, client interaction, and activity selection—the nuts and bolts of practice. It does not attempt to be a comprehensive work on the field of horticultural therapy. Horticultural therapy also requires skills in program management, development, funding, garden design, enabling garden techniques, and facility management. More .in-depth education and experience in understanding and serving populations encountered in horticultural therapy programs are indispensable to maximize effectiveness. See the "Core Curriculum" of the American Horticultural Therapy Association (www.ahta.org) for a listing of these and other recommended topics for the continued study of horticultural therapy.

Readers should recognize that differing views exist within the profession on many of the topics presented. The perspectives given in this text represent each author's viewpoint and are not intended to represent the official views of the American Horticultural Therapy Association or any other professional organization.

The editors hope that the information in this book will boost the use of horticulture for therapy, rehabilitation, and wellness to continue to evolve in practice, scope, and recognition—so that more people may experience its benefits.

Acknowledgments

We are grateful to the many people who contributed to the production of this book, notably the authors (Pam Catlin, Karen Kennedy, and Sarah Sieradzki) who provided their incredible experience and understanding of horticultural therapy; Jay Rice, for his thoughtful contributions to "Horticultural Therapy Practice" throughout the book; and Heather Benson, who researched and compiled horticultural therapy techniques.

Thank you to my son, Shawn Cremer, for sharing an enthusiastic relationship with nature, and to Moss Cremer for his support and patience. Thank you to Christine Kramer for having the guts, resolve, and skills to work with me for so long—first in horticultural therapy at Denver Botanic Gardens, later as we began the Horticultural Therapy Institute in 2002, and most recently on this book. Without her encouragement and organization, we could not have pulled it off. I am also grateful to the students of horticultural therapy whom I have had the pleasure to teach over the years. They are an incredibly strong and determined lot. Thank you also to the late Professor Richard H. Mattson, who pushed me to work with people with developmental disabilities and was my mentor in horticultural therapy; to Bruce Christensen, who first hired me as a horticultural therapist and inspired me to respect the people in our programs; and to the staff at Mountain Valley Developmental Services and Big Lakes Development Center for teaching me so much. Finally, my thanks go out to the individuals whom I have served in horticultural therapy programs.

Rebecca L. Haller

I met Rebecca Haller in 1994 and continue to be impressed by her dedication and drive to spread the gospel of horticultural therapy. Always uppermost in her mind is how to provide her students with excellence in horticultural therapy education. I believe she has been successful, and this book is just an extension of that strong commitment to help create leaders in this field. It's been my pleasure to learn from and work with such a devoted individual. I would also like to thank my former editor at the *Denver Catholic Register*, James Fielder. He was the finest editor I've ever had the privilege of working with in this field. Bless you, Jim. Last, I acknowledge my parents, Josephine and Jerry Capra, for always believing in my abilities, and my greatest cheerleader—my grandmother, Angelina Durando.

Christine L. Capra

Editors

Rebecca L. Haller has practiced and taught horticultural therapy since receiving an MS in horticultural therapy from Kansas State University in 1978. Currently the director of the Horticultural Therapy Institute in Denver, Colorado, Ms. Haller delivers workshops, teaches horticultural therapy classes in affiliation with Colorado State University, and provides consultation to new or developing programs. At Denver Botanic Gardens, she designed and taught a series of professional courses in horticultural therapy, managed the sensory garden, and created programs and access for people with disabilities. She established a vocational horticultural therapy program for adults with developmental disabilities in Glenwood Springs, Colorado, which is still thriving after more than twenty years in operation. She has served as president, secretary, and board member of the American Horticultural Therapy Association (AHTA), and has worked on teams related to education and professional standards. In 2005, she received the Horticultural Therapy Award from the American Horticultural Society. Awards from the American Horticultural Therapy Association include: the Publication Award in 2009 and the Rhea McCandliss Professional Service Award in 2015. She has authored articles and chapters in the following publications: *HortTechnology, Horticulture as Therapy: Principles and Practice, Horticultural Therapy and the Older Adult Population, Towards a New Millennium in People-Plant Relationships, AHTA News Magazine, and The Public Garden.*

Christine L. Capra is the program manager at the Horticultural Therapy Institute in Denver, Colorado. She and partner, Rebecca L. Haller, HTM, cofounded the institute in early 2002 and continue to offer accredited horticultural therapy education to students from across the United States and abroad. Previously, she worked as the program coordinator at the Denver Botanic Garden's horticultural therapy program. She has a BA in journalism from Metropolitan State University in Denver. Ms. Capra has written for numerous publications including *OT Weekly, Mountain Plain and Garden, Green Thumb News, People-Plant Connection, AHTA News, GrowthPoint, The Community Gardener, Health and Gardens, Colorado Gardener, Our Sunday Visitor,* and *Jardins.* She was a reporter for the *Denver Catholic Register* newspaper for many years. She has received writing awards from the Society of Professional Journalists, the Catholic Press Association, and the Colorado Press Associations. In 2009, she was awarded the Publication Award from AHTA.

Authors

Pamela Catlin has worked in the field of horticultural therapy since 1976 and has been instrumental in establishing programs in Arizona, Illinois, Oregon, and Washington. She is currently the director of horticultural therapy with Adult Care Services, providing programming to adults with Alzheimer's and other forms of memory loss at the Margaret T. Morris Center. She also serves adults with physical, mental, and emotional challenges at the Susan J. Rheem Adult Day Centers in Prescott and Prescott Valley, Arizona. Pamela is on the faculty of the Horticultural Therapy Institute, serves as a Prescott College mentor and instructor, supervises interns, and does private contracting in horticultural therapy. She has authored chapters in the textbooks *Horticulture as Therapy: Principles and Practice* (Simson & Straus) and *Horticultural Therapy Methods: Making Connections in Health Care, Human Service, and Community Programs* (Haller & Kramer). She wrote and published *The Growing Difference: Natural Success Through Horticultural-Based Programming*. Pamela has been a speaker at conferences for (AHTA), American Therapeutic Recreation Association, National Adult Day Services Association, Pioneer Network and more.

Karen L. Kennedy has been active in the field of horticultural therapy since 1986, developing programs and providing services to individuals with a wide variety of disabilities, illness, and life situations. After managing the Horticultural Therapy and Wellness Program at The Holden Arboretum for 23 years, Karen now works as a private contractor providing horticultural therapy and consulting services, developing educational materials, and teaching. She is on the faculty of the Horticultural Therapy Institute, Denver, Colorado. In addition, she nurtures her love of plants as the Education Coordinator for The Herb Society of America. She has presented at regional and national professional conferences and served on the board of directors and committees of AHTA. Karen enjoys writing and has coauthored chapters in the textbooks *Horticulture as Therapy: Principles and Practice*, (Simson and Straus), *Horticultural Therapy Methods: Making Connections in Health Care, Human Service, and Community Programs* (Haller and Kramer), and an appendix on HT in *Public Garden Management*

(Lee). Karen received the 1994 Rhea McCandliss Professional Service Award from AHTA and 2009 Horticultural Therapy Award from the American Horticulture Society. She holds a B.S. in horticultural therapy from Kansas State University.

Sarah G. Sieradzki graduated from Indiana University in 1976 with a B.S. in occupational therapy. She worked in special education, has specialized in mental health practice since 1986, and is currently a clinical specialist in mental health occupational therapy at University Hospitals Case Medical Center in Cleveland, Ohio. Beginning in 1992, Sarah actively pursued education and experience in horticultural therapy and has spoken about HT at local, state, and national conferences. She co-led a horticultural therapy support group at Holden Arboretum, completed the certificate program at the Horticultural Therapy Institute in Denver, and became a registered horticultural therapist in 2008. Since 2012, Sarah has been teaching at Cleveland State University Masters of Occupational Therapy program, and at the Physical Therapy Assistant program at Cuyahoga Community College in Cleveland. She has also served as a board member of the Michigan Horticultural Therapy Association since 2011.

Contributors

Heather G. Benson is currently an educator, a horticulturalist, and manager of therapeutic horticulture programming for the University of Minnesota. She provides services to the following populations: special education students and adults with developmental disabilities, as well as people with eating disorders, Parkinson's disease, and dementia. Before returning to Minnesota in 2013, Heather delivered therapeutic programming for a wide range of populations in urban settings in the San Francisco Bay Area. She coordinated a public school garden program, offered therapeutic horticulture to special education students, and managed after-school programs in community gardens.

Jay Stone Rice is on the faculty of the Horticultural Therapy Institute. He was the principal investigator for an exploratory study of the effectiveness of the San Francisco Sheriff's Department's innovative Garden Project and the impact of early trauma and loss on the county jail inmate population. Dr. Rice has written about the social ecology of inner city family trauma, the relationship of trauma to substance abuse and crime, gardening as a treatment intervention, the neurobiology of people–plant relations, and understanding our human nature. He has consulted on the development of ecologically sensitive treatment programs and is a family therapist practicing in San Rafael, California.

chapter one

The framework

Rebecca L. Haller

Contents

Introduction

> Whatever it is that calls the gardener to the garden,
> it is strong, primeval and infinitely rewarding.
>
> **Lauren Springer**
> *The Undaunted Gardener, 1994*

Horticultural therapy offers positive and rewarding experiences for program participants, therapists, and those who come into contact with the growing environment. People involved in these programs intuitively know the many benefits and joys derived from connecting with nature (Figure 1.1). The attraction to engage in horticultural therapy activities may stem from a deeper "pull," in addition to the visible positive effects. Beneath the surface of this life-enhancing practice lie the conscious steps that are employed by the horticultural therapist to provide therapeutic programs. As an emerging profession, horticultural therapy continues to utilize the techniques practiced by many related health and human service fields—notably psychology, occupational therapy, vocational rehabilitation, social work, therapeutic recreation, and education. Much has been written about the reasoning and processes that are employed in these and

Figure 1.1 Gardening offers a nature connection. Photo by Rebecca Haller.

other fields of human service. Theoretical bases for practice vary by setting, population, and therapeutic approach. However, the basic processes used are remarkably similar across disciplines. Horticultural therapy practitioners who adopt these accepted treatment procedures are able to positively impact the quality of their services and the profession as a whole.

This chapter provides a framework for discussion of the processes and techniques outlined in later sections of this manual. Included are systems and events that shape the horticultural therapy profession, a working definition of horticultural therapy, an overview of program types and the people served, and reflections on the significance of horticulture as a therapeutic medium.

Shaping a profession

The practice of horticultural therapy has progressed from an 1800s belief that working in the agricultural fields could benefit mental patients, to the use of gardening as activity and therapy for physical rehabilitation in the early 1900s, to the presence of many types of programming and settings in the 2000s. Entities that have influenced—and continue to shape—the profession of horticultural therapy include practitioners, educators, researchers, professional associations, volunteers, regulators, and employers, as well as clients or participants.

Practitioners

Since the mid-1900s, horticultural therapy has been used by mental health professionals, occupational therapists, physical rehabilitation specialists, and vocational service providers, as well as other human service providers. In some settings, horticultural therapists work closely, or co-treat, with allied professionals to maximize outcomes and benefits to clients. The profession is characterized by a prevalence of practitioners who are willing to share information freely and are open to new ideas and approaches. A healthy variety of program models results largely from the creativity and dedication of these diverse professionals.

Educators

Horticultural therapy education and training programs have led efforts to produce research, helped to define the profession through curricula content, and have been well-represented in establishing credentialing standards. Historically, and currently, most educational programs are based in horticulture or plant science departments of colleges and universities, or in horticultural therapy certificate programs. Seldom are they found in human science or health-care facilities. Future curricula with a balanced proportion of human services and social science courses are needed to reflect the

Figure 1.2 Horticultural therapy education helps to shape the profession through curriculum and training. Photo by Christine Capra, Horticultural Therapy Institute.

interdisciplinary competencies required of horticultural therapists who wish to practice in health-care or social service arenas (Starling et al. 2014) (Figure 1.2).

Researchers

Many benefits of horticultural therapy have been portrayed anecdotally in publications over the past 50 years. Research that documents outcomes and efficacy also exists but is much less plentiful. The need for strong research as a base for the profession cannot be overstated. Practitioners, educators, and researchers from relevant disciplines must team together to carry out and publish research that employs sound social science methods. This is critical for future funding, employment, and excellence in practice.

Professional associations

In the United States, the American Horticultural Therapy Association (AHTA) and its regional networking groups have focused on information dissemination through publications, conferences, and networking. In order to advance the practice of horticultural therapy and expand employment options, future efforts must include strong advocacy and the promotion of the profession to health-care and human services providers as well as regulatory and insurance leaders (Haller 2003).

Another important activity of AHTA is to create and manage a credentialing system—currently a voluntary professional registration system based on interdisciplinary education and a supervised internship. This system has recognized the scarcity of university degrees in horticultural therapy and has formerly given credit for a wide array of education and experiences. AHTA plans that future standards will include a certification exam based on job knowledge and skills (Starling et al. 2014).

Similar associations exist across the globe, including in Canada, Japan, Australia, the U.K., and others.

Volunteers

Since the mid-1900s, volunteers have brought gardening activities to residents of prisons, hospitals, long-term care, and others—usually at no cost. Garden club members and master gardeners have been especially active in developing programs utilizing their training and experience in horticulture. Health-care or human services education is seldom required, resulting in programs of varied therapeutic value. Distinguishing between horticultural therapy programs and articulating a monetary value for volunteer services can help to eliminate the confusion that exists among administrators and potential employers. When volunteers serve as resources for programs led by trained horticultural therapists, optimum conditions for sustainability and effectiveness exist.

Regulators and employers

Employers commonly look to supplemental funding bases to operate horticultural therapy programs. Private donations of money, materials and labor, self-earned income from sales of plant products, and program grants are frequently vital to the provision of program funding. In health care, insurance companies regard horticultural therapy services as reimbursable only when they are framed within strict guidelines such as for training or co-treatment.

In order for the practice of horticultural therapy to be regarded as effective and fundable by administrators, insurance companies, and regulators, the following nationally coordinated actions must be taken: build a strong research base, apply standard treatment procedures to practice, develop a rigorous credentialing system, and advocate for the profession.

Clients/participants

The people serving in horticultural therapy programs play a role in how programs evolve. With a trend toward client-centered care, those involved may direct treatment plans and choose therapies, placements, and pursuits in which to be engaged. As more consumers understand the benefits of using horticulture as a tool for therapy and rehabilitation, they are more likely to choose the organizations that offer this service. For example, an individual may prefer to live in an assisted-living or long-term care facility that has a gardening program as an option for ongoing activity.

This may give the organization a competitive edge and the motivation to continue or expand horticultural therapy programming.

Horticultural therapy defined

As might be expected for a relatively young profession with diverse applications, a full spectrum of published definitions of horticultural therapy exists. Definitions encompass strict portrayals of horticultural therapy in health-care terminology as well as those that broadly include any beneficial horticultural experience (Dorn and Relf 1995). Recent authors have generally defined horticultural therapy more narrowly while using alternate terminology for the positive effects that gardening and passive garden exposure can have on the general population (Matsuo 1992, Sempik et al. 2003, Diehl 2007). In 2007, the AHTA described four types of horticultural therapy programs, including horticultural therapy, therapeutic horticulture, social horticulture, and vocational horticulture (Diehl 2007). Some horticultural therapists now refer to their programs as therapeutic horticulture, particularly those that include active and/or passive involvement with plant-related activities that do not focus on clinically defined treatment goals. Thrive, an organization in the U.K., uses the terms "social and therapeutic horticulture" to describe a full range of program types and purposes for health and well-being (Thrive 2015).

The following definition is also presented to describe horticultural therapy in its many forms:

> Horticultural therapy is a professionally conducted client-centered treatment modality that utilizes horticulture activities to meet specific therapeutic or rehabilitative goals of its participants. The focus is to maximize social, cognitive, physical and/or psychological functioning, and/or to enhance general health and wellness.

This definition includes three elements as described by Dorn and Relf (1995): the clients served, therapy goals, and treatment activities. They describe horticultural therapy as a practice that serves *defined client groups* (those with identified therapeutic or rehabilitation needs), that is *goal driven* (based on standard treatment procedures), and that uses the *cultivation of plants* as its primary treatment activities. These authors state the importance of the presence of all three elements to distinguish horticultural therapy from other types of garden interactions. In a graphic model of horticultural therapy, Mattson depicts the various interactions of a *client, horticultural therapist,* and *plant* during a horticultural therapy session (Mattson 1982). A key element in this model, the trained horti-cultural therapist has the skills to use plants to facilitate the therapeutic process.

An adaptation of these earlier models shows the client as the central figure within the process of a horticultural therapy interaction (Figure 1.3). The client is both the receiver and initiator of the treatment process. In this model, there are four elements, including *client, goals, therapist,* and *plant.* The *client* is the person being served, usually someone with an identified need for intervention to improve cognitive, emotional, physical, or social functioning. *Goals* are those treatment goals and objectives defined by the client and the treatment team. The *therapist* is the professional who is trained in the use of horticulture as a modality for therapy, rehabilitation, and wellness—the horticultural therapist. The term *plant* is used here to signify those garden and plant-related activities and tasks used to provide therapeutic opportunities to the client. Pathways lead to and from each of the four elements. The therapist interacts with the client through the plant or by direct contact. Plant activities are chosen to meet the goals of the client. The client interacts with all other elements and is in the center of the diagram to illustrate a focus on the participant in the horticultural therapy treatment process.

Note that horticultural therapy is based on purposeful horticultural activity or work—the cultivation of plants. Passive enjoyment of gardens and plants may be included as part of a program, but not to the exclusion of this active participation. Similarly, the guidance, judgment, and cre-ativity of a trained therapist and stated treatment objectives are essential elements in horticultural therapy programming in order to maximize the benefits derived from the people–plant interaction.

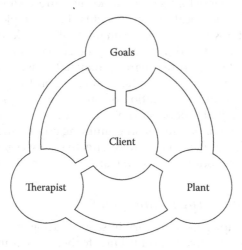

Figure 1.3 Horticultural therapy practice: elements and process.

> **Exhibit 1.1   Horticultural Therapy Practice:**
> **Community Horticulture**
>
> Within this manual, the term *community horticulture* is used to describe those programs that use horticulture to improve quality of life through the development of individuals and communities. Examples of sites include community gardens and greenhouses, school gardens, urban greening sites, nature centers, healing gardens, and rural development projects. Like horticultural therapy programs, the purpose is to offer services that provide human benefits including cognitive, psychological, physical, and social gains. In practice, the boundaries between these programs as well as between therapeutic horticulture and horticultural therapy are often blurred.

It is anticipated that the definition presented here will help practitioners to know when and if their work should be identified as horticultural therapy. Programs that offer gardening activities without goals and treatment procedures, while beneficial, are not generally considered horticultural therapy. As an active participant in the horticultural therapy process, the therapist is responsible for planning treatment, developing sessions, interacting with the client, and recording outcomes or results.

Techniques and skills related to each of these elements are found throughout this manual. The training materials presented are based on the use of horticultural therapy as a treatment modality as defined in this section, with the intention that they will also enhance the skills of those who use horticulture as a tool for therapeutic horticulture, community development, and other nonclinical human benefits. For example, a community garden coordinator who runs a program for youth at risk will benefit from writing clear objectives and documenting outcomes. Although these and other community horticulture programs may not be within a strict definition of horticultural therapy, they have in common the focus on human benefits rather than plant cultivation or physiology. Employing the basic methods described in this book will help community horticulture program leaders to clarify intentions, focus programming efforts, and produce measurable outcomes. In addition to more effective services, results include potential for new or continued funding and other support (Exhibit 1.1).

Program settings, types, and goals

Horticultural therapy programs vary in scope, setting, purpose, and treatment approaches. This diversity contributes to their appeal and affordability. From a few plants on a windowsill used for infrequent sessions to full-time programming and year-round growing facilities, programs can be customized to fit the needs and resources of each organization employing them.

Settings

Horticultural therapy programs are involved in health care, human service, health promotion or wellness, and community development. Settings include rehabilitation hospitals, mental health facilities, vocational training centers for people with developmental or other disabilities, correctional institutions, long-term care and assisted-living facilities, schools, and community gardens. The purpose of programs can vary even within a setting. For example, in a rehabilitation setting, a horticultural therapy program may assist patients to regain functional physical or cognitive skills, help patients emotionally cope with loss or life-altering disabilities, provide guidance and exposure to a new leisure activity to increase quality of life, and/or may serve as a forum for improving social interactions and relationships. In a school setting, developmentally disabled students may participate in horticultural therapy sessions to improve work habits, social skills, or cognitive processes, whereas children with behavioral challenges may garden in order to learn socially appropriate behaviors, communication skills, and stress management strategies.

Types

Programs may be categorized into three types (Haller 1998): vocational, therapeutic, and social. Vocational programs seek to affect vocational outcomes of participants, improving job proficiencies and employment opportunities. Rehabilitation or habilitation of the wide span of skills that impact success in a work environment are typically addressed, including those in cognitive, physical, and psychosocial areas. Many programs may employ a task-oriented approach to build and enhance abilities for successful engagement in the community as well as the work setting. A task-oriented approach might also be found in therapeutic programs where the focus is on recovery from mental or physical illness or injury. Also addressing broad and whole-person issues, therapeutic programs are based on a medical model and seek to optimize physical and mental health. A third type of horticultural therapy program, social or wellness, improves general health and/or quality of life of participants. Programs are holistic and typically self-directed to address some or all of the dimensions of wellness such as occupational, physical, social, intellectual, spiritual, and emotional facets of life (National Wellness Institute 2015). All three of these program types support individuals in the growing process to achieve their full potential. (See Exhibit 1.2 for an example of choosing a program type to offer.)

**Exhibit 1.2   Horticultural Therapy
Practice: Finding the Right Niche**

When I was approached to set up a new horticultural therapy program for adults with traumatic brain injury in a prevocational day treatment setting, my first task was to determine the niche or role horticultural therapy would play in the overall program. Horticultural therapy could provide a wellness element to facilitate self-care, address coping skills for managing recovery, be a creative outlet for self-expression, or provide an opportunity for working on vocational skills.

Meeting with the program director helped me understand her vision for the program, participant demographics, average length of stay, and the next steps for participants "graduating" from the program. She shared that the core program components included daily living skills, personal adjustment training, health and wellness, and vocational skills.

After observing group sessions already in place, we decided that the horticultural therapy component would have a vocational focus. Operating like a small business, participants selected themes and grew plants for several plant sales annually. Planning, organizing, and budgeting were among the tasks. Participant goals and objectives focused on basic work skills including commitment to quality standards, endurance, working with supervisors, getting along with coworkers, and developing compensatory strategies. Participant progress was reviewed monthly at team meetings comprising all the professionals working in the program.

Contributed by K. Kennedy

Goals

Within each of these program types, participants may work on specific goals in some or all of the developmental areas that are potentially impacted by horticultural therapy services including cognitive (intellectual), emotional, social, and physical (Olszowy 1978). The emphasis varies depending on the program type and the individual clients served. Horticultural therapy is concerned with the needs and goals of the individual and whole person rather than with a particular disease, diagnosis, or disability.

Horticultural therapy utilizes horticulture activities to facilitate change in program participants. The purpose and focus of these activities varies by program type and by specific treatment objectives. Hagedorn

(1995) identifies five focuses for the "applied use of activities." They are *focus on product, process, competent performance,* the *individual interacting with others,* and the *individual interacting with the environment.* Each of these approaches has merit and usefulness in horticultural therapy practice.

A *focus on product* or the end result of the activity provides meaning and motivation to the participant. That a well-grown window box or harvest of cherry tomatoes is seen as a worthwhile product results in therapeutic benefits to the client. The client is willing to participate and value the experience.

Therapeutic goals may be achieved by a *focus on the process* of doing an activity. Commonly used in horticultural therapy programs, the process itself can elicit improved mood or attention. By engaging in the process of gardening, a participant may experience flow (Hagedorn 1995) or a sense of fascination (Kaplan and Kaplan 1989) in which attention is effortless and complete, resulting in recovery from mental fatigue.

Through *competent performance* of an activity, the individual can begin to improve self-concepts and break negative cycles of real or perceived abilities and control. These inner changes can be catalysts for other therapeutic and developmental improvements. Through involvement in horticulture activities, a seed of change may be planted. For example, a patient may become more receptive to other therapeutic interventions and make progress in diverse areas. An inmate may perceive that he or she can succeed in school or work, leading to self-improvement in areas beyond the horticulture program.

Horticultural activities have the potential to be highly effective means for improving social *interaction.* Activities can be structured to encourage cooperation and communication among participants as well as to build relationships between the client and therapist. Barriers to communication are effortlessly reduced through shared garden work as shown in Exhibit 1.3.

Through activity, the participant *interacts with his or her environment* and consequently changes it in some way. In the horticulture environment, the individual is also positively affected by this interaction with nature, with opportunities for growth, restoration, and enjoyment (Kaplan and Kaplan 1989, Neuberger 2008).

More detailed descriptions of horticultural therapy settings, program types, and goals may be found in *Horticulture as Therapy: Principles and Practice* (Simson and Straus 1998).

Understanding the people served

Understanding the populations who are served by horticultural therapy is essential in order to plan and implement appropriate and effective treatment programs. While it is important to know the cause, effect,

---

**Exhibit 1.3   Horticultural Therapy Practice:
Facilitating Change**

Goals to improve social interaction are often addressed during group activities, for example:

- *Goals*: To improve communication skills, including making eye contact when speaking and initiating conversation.
- *Task*: To create a new summer garden.
- *Process*: The group meets to discuss the appropriate location for the garden, plants to include, and steps needed to prepare the site. The steps to prepare the site as well as the planting are divided and shared among the participants. This could occur over several sessions, depending on the time frame and the level of the group. Each step in the process of creating a new garden offers many opportunities for participants to address their communication skill goals.

Contributed by K. Kennedy

---

and progression of many types of health conditions and disabilities, it is especially useful to recognize how conditions affect the individual's functioning and overall wellness. This knowledge impacts the reasoning, decisions, and interactions of the therapist on a daily basis.

For treatment planning, the horticultural therapist must know enough about a particular type of illness or disability, and most importantly, about the individual, to develop activities or adaptations that are appropriate and accessible to the participant. Similarly, the organization or structure of the activity space must be based on the needs and abilities of those served. This allows the client to access the space and the activity as well as experience appropriate challenges and problem-solving opportunities.

Safety considerations are also based on knowledge of the physical and cognitive skills and limitations of participants. The horticulture medium presents potential safety hazards through the use of tools, physical exertion, exposure to weather and sun, the cultivation of toxic plants, and the use of toxic or dangerous program materials. Informed choices must be made regarding what are or are not acceptable risks for each person served. In many cases, creativity is required to reduce hazards. (See Table 1.1 for some examples of solutions to safety issues for an elder with memory impairment.)

Interactions between the therapist and client are improved if the therapist is familiar with key characteristics that influence communication

Table 1.1 Horticultural therapy practice: Safety precautions

Issue	Solution
Susceptible to sunburn	• Ensure that the participant has sunscreen, long sleeves, hat, and sunglasses • Schedule sessions when garden is not in full sun, or work in shaded areas if possible
May ingest nonfood items	• Use nontoxic plants and program supplies
Shuffles when walks	• Use hose reel to keep hose off walkways • Use paved or hard-packed surfaces
Disorientation or confusion	• Create pathways in a circular pattern with no dead ends
Impaired judgment	• Limit access to sharp tools

Source: Contributed by K. Kennedy.

Note: Safety precautions vary with program settings and needs of the participants so it is important to think through the issues and plan creative solutions.

with a particular population. For example, it is effective to use simple and adult-appropriate language with an adult with an intellectual disability. For someone with a hearing impairment, the therapist must speak clearly and look directly at the participant. In communicating with a person who has experienced a traumatic brain injury, it may be useful to provide written or pictorial directions to augment verbal instructions.

Practical descriptions of client populations may be found in "Part Two: Special Populations for Horticultural Therapy Practice" of *Horticulture as Therapy: Principles and Practice* (Simson and Straus 1998). For information on diagnostic groupings, symptoms, functional limitations, and implications for treatment, see *Therapeutic Recreation and the Nature of Disabilities* (Mobily and MacNeil 2002). Many of the major populations encountered in horticultural therapy are described in *PocketGuide to Assessment in Occupational Therapy* (Paul and Orchanian 2003). Included are physical, psychosocial, geriatric, pediatric, and developmental diagnoses, with summaries addressing etiology, symptoms, and outcomes for each condition.

In addition to this information, therapists need to understand the experience and social systems of the people served, as described in Exhibit 1.4. Background information on general populations, combined with empathy and specific knowledge about each individual, are essential for effective horticultural therapy programming.

Horticulture as a therapeutic modality

The use of horticulture as a therapeutic modality (or medium) offers numerous advantages, including a universal appeal, flexibility, and wide-ranging impact. What makes this medium special?

Exhibit 1.4   Horticultural Therapy Practice:
Understanding the People We Serve—An
Ecological, Empathetic Vantage Point

Heinz Kohut, the originator of self-psychology, suggested that empathy provides us with the experience of ourselves as a whole and vital self (Baker and Baker 1987, Kohut et al. 1984). He defined empathy as the capacity to understand another's experience as well as the ability to mirror their experience back to them. Think of a time when someone you knew noted some quality that he or she perceived in you. While you may have never thought of yourself in this way, you immediately felt the truth of what was said to you. In these moments of mirroring, we develop a fuller view of ourselves as well as feeling more significant and valued because we have been seen. Kohut suggests that, throughout our lives, our psychological well-being is enhanced by our empathic connections to others.

Empathy requires the ability to reflect upon our own life experience and to imagine the experience of others. In *Number Our Days*, Barbara Myerhoff, the noted cultural anthropologist, describes imaginative identification, a process she utilized in studying aging among the members of a Jewish senior center in Venice, California.

> At various times, I consciously tried to heighten my
> awareness of the physical feeling state of the elderly by
> wearing stiff garden gloves to perform ordinary tasks,
> taking off my glasses, plugging my ears, slowing down
> my movements and sometimes by wearing the heavi-
> est shoes I could find to the Center. ... Once by accident
> I stumbled slightly. The flash of terror I experienced was
> shocking. (Myerhoff 1978)

In *The Ecology of Human Development*, Urie Bronfenbrenner suggests that empathy is aided by looking at the systems within which the individual develops as well as those that currently impact the person's life (Bronfenbrenner 1979). How did they grow up, who was available to meet their needs within their families, schools, churches, or communities? What was the level of communication between the family and others in their community? Were they isolated, alienated, or supported? If their parents worked, were they able to take time off to meet the needs of their children? What was the area like in which they grew up, physically, economically, and socially? How does the larger community view them? What private or public resources are committed to support or aid their participation in the community? When we work with someone who is disabled, aging, or incarcerated, we may not get this information directly from him or her. However, secondary sources such as research studies, books, and films can help us begin to sense what it is like to walk in their shoes.

Contributed by J. S. Rice

Encourages human growth

Charles A. Lewis devoted much of his life to understanding and communicating the positive effects of gardening on the gardener. In *Green Nature/Human Nature* (Lewis 1996), he described several characteristics of gardening that encourage human growth and wellness. These include:

- Enhanced pride and self-esteem resulting from the physical beauty created
- Deep personal involvement because individuals tend to become deeply engaged in the gardening process
- Requirements of patience and delayed gratification
- Awareness of natural forces and rhythms
- Interdependence or partnership between gardeners and the plants they tend, forming a symbiotic relationship whereby the plants get tended and the gardener experiences a sense of purpose
- Inner peace based on the natural rhythms and dynamic stability of gardens, in contrast to modern schedules, fads, and threats
- Opportunities for focused attention that provide rest from mental fatigue and worry

Each of these innate or intrinsic characteristics of gardening lends something unique to the therapy experience (Figure 1.4).

Offers restoration

Rachel and Stephen Kaplan (Kaplan and Kaplan 1989) report that natural environments, including gardens, are restorative and are consistently preferred over many other environments. Restoration or "recovery from mental fatigue" results from contact with nature. Additionally, gardener-reported experiences of "enjoyment, relaxation, and lowered stress levels" may ultimately affect physical health. Natural environments may also distract and allow the person to focus on something other than his or her current problem (Marcus and Sachs 2014).

Figure 1.4 Gardening encourages human growth and wellness. Photo by Anna Terceira.

Addresses innate psychological needs

Many authors have noted that gardening taps into a universal and innate psychological need for connections with plants and the earth (Kaplan and Kaplan 1989, Ulrich 1993). Additionally, gardening can provide many benefits:

- Mental growth, brain development and function (see Exhibit 1.5 Horticultural Therapy Practice: Horticultural Therapy and the Human Brain)
- Psychological benefits such as tranquility and healing and emotional health
- Improved human relationships and communication
- Physiological conditioning
- Money from selling produce or saving on food costs
- Improved surroundings or environment
- Food and other useful products
- Education in diverse topics

Exhibit 1.5   Horticultural Therapy Practice:
Horticultural Therapy and the Human Brain

Horticultural therapy may be uniquely suited to supporting balanced brain functioning and facilitating our capacity to grow and change in response to life challenges (Rice 2012). Paul MacLean (1973) proposed an evolutionary, explanatory model for understanding our brain functioning in his Triune Brain Theory. He formulated that the brain comprises three sub-brains, developed to meet specific conditions during the course of evolution. The reptilian brain at the top of the spinal cord maintains the heartbeat, breathing, and the ability to swallow. This sub-brain also activates the startle reflex, which enables us to respond to perceived threat by stimulating the production of cortisol and adrenalin. This sub-brain's primary focus is our survival.

Charles Darwin observed how the evolutionary emergence of mammals requiring extended nurturance post birth stimulated the development of the limbic brain (Darwin 1998). This sub-brain extends our nervous system's capacity to perceive and meet the needs of someone outside of our self. The limbic system shapes and expresses our emotional life. Attunement to others, attachment, security, and emotional intelligence are the domains of this sub-brain. The experience of nursing, intimacy, and touch, among other things, stimulates the production of the neuropeptide, oxytocin. This important neurochemical plays a role in neuroplasticity, the brain's capacity to grow new synaptic connections, which is critical for adapting to changing conditions over the course of the human lifespan.

The neocortex, the last sub-brain to develop, provides us with the capacity for abstract reasoning, planning, and perception. The neocortex enables humans to think of ideals that exist outside of nature. This capacity has spurred great technological advances. Yet, as Albert Einstein (1995) has written, it is important not to deify our intellect. For without integration with the other sub-brains, the neocortex can distort our understanding of our lives and of life itself.

When we face sudden change, loss, or trauma, we have a myriad of feelings and reactions. We may experience fear generated by our reptilian brain's concern for our survival. We may experience a loss of identity, meaning, and purpose, as the previous ideas about our lives, generated by the neocortex, are challenged and disrupted. We long for a sense of connection that will carry us through the turmoil of transition. Participating in the care of plants promotes attunement with the larger cycles of nature. Stephen Kaplan (1978) observed how engaging with nature promotes the reflection necessary for adaption and survival. Horticultural therapy, by fostering a relationship with plants, may stimulate the production of oxytocin, and thereby support our regaining our balance through engagement with our human nature and the natural environment that sustains us.

Contributed by J. S. Rice

Offers versatility

The medium of horticulture is a flexible environment that offers opportunities that cross cultures, ages, socioeconomic conditions, physical/mental/social/emotional conditions, and health status. Activities range from the many aspects of planning, preparing, planting, cultivating, and harvesting to using products and recycling garden wastes. The majority of tasks are easily adaptable to accommodate the abilities and challenges of diverse individuals.

Has meaning and purpose

Another key benefit of using horticulture is that it is a meaningful, purposeful activity that is motivating, normal, and tangible. In other words, it is "real." Many people enjoy the process as part of a normal, healthy lifestyle (Matsuo 1992). This normalcy and realness helps break down the barriers experienced by many program participants. Gardening is common to many people's experiences and family history. It also transcends barriers established by socioeconomic status, language ability, and disabilities. It produces a product of value to others and enables the program participants to become the caregivers and the providers (see Table 1.2).

Table 1.2 Horticultural therapy practice: Meaning and purpose

Individual	Meaning
Older adult man in a nursing home	Gardening is a gender- and culturally appropriate activity and a continuation of a previous hobby.
Adult with intellectual or developmental disabilities in a group home	Gardening provides produce for the individual and staff as well as an activity and topic of conversation in common with community neighbors.
Adult with mental health issues in an outpatient clinic	Produce stand provides vocational and social skills by providing a valued commodity and opportunity to interact with the community.

Source: Contributed by K. Kennedy.

Figure 1.5 Gardeners connect with the natural world. Photo by Rebecca Haller.

Gardening is a process that allows the gardener to be part of something bigger, to connect with nature, the community, and life. Horticultural therapy can provide experiences and insight to enhance this sense of connection with natural rhythms (Figure 1.5). (See Exhibit 1.6 for further information and examples of using nature ceremonies in horticultural therapy programs.)

Impacts others

Note that, perhaps more than any other therapeutic medium, horticulture affects even those who do not receive horticultural therapy services. All who come in contact with horticultural therapy gardens and the resulting facility improvements can benefit from this passive interaction. This applies to visitors, administrators, and staff as well as residents who experience the contact with nearby nature on a daily basis.

Considering the importance of the horticultural environment and the role of plants in healing as described in Exhibit 1.7, it is imperative that horticultural therapists have a solid understanding of this medium as well as the necessary therapeutic skills to provide treatment. In horticulture, skills

Exhibit 1.6　Horticultural Therapy Practice: Working with Nature: Ceremonies for Growth and Understanding

When I first moved to San Francisco, I came into a relationship with a 1968 VW bus. One November day, I decided to try rebuilding the engine on my own. I bought a copy of *How to Keep Your Volkswagen Alive* (Muir 1969) and arranged to use my friend's driveway. November in the Bay Area happens to be the start of the rainy season. Over the course of the next two months, I spent much time on my back on wet and cold concrete. On a cellular level, I was learning that it is best to make plans in harmony with the seasons. Farmers, gardeners, and horticultural therapists take direction from natural cycles as a matter of course. It would be easy for the horticultural therapist to take the aligning of oneself with nature for granted and miss how vital an offering this is to many people growing up in our modern and postmodern era. Humans, as part of the natural world, suffer when they do not recognize how they are affected by natural cycles.

A nature ceremony is one that aligns us with the season. For example, many people have difficulty with darker moods during the cold winter months. This condition is sometimes severe enough to be given the diagnosis of seasonal affective disorder, otherwise known as SAD (American Psychiatric Association 2000). One treatment for this condition is exposure to light. Perhaps this is why Alaska Airlines has so many flights to Mexico in the winter. We might consider if there is something to be gained by the dark moods we experience during the winter months. After all, if we look at a deciduous tree in the middle of December, it would be easy to diagnose it with depression. It has no fruits or leaves and looks close to death. However, we know that the tree's response to the season is to draw its life force to its roots deep within the earth. Here, it conserves energy and prepares for renewal in the spring. Perhaps human suffering at this time of year is in part because we often try to live our lives in the winter as if it were summer. We do not slow down and honor the pull of nature; rather we often push against it.

A nature ceremony teaches us that life is a process rather than a series of activities that are judged by whether they bring us immediately to our goals. Rabbi Gershon Winkler wonders if the apple is disappointed when its seed produces a stick (Winkler 2005). In nature, we learn patience by observing the stages of creation that precede and follow fruition. In a nature ceremony, we can help those we work with locate where they are in the cycle of an activity or their life journey (Eagle Star Blanket 2004). We can plot out the stages: seed, germination, planting, new starts, flowering, broadcasting, death, compost, and awaiting rebirth. Through ceremonies, we learn how to locate our human experience through metaphorical reflection and actual experience of our natural life cycle.

There are many forms of nature ceremonies that horticultural therapists can utilize. You can ask the person you are working with to plant an intention along with the planting of a seed. You can ask them to think of something they may wish to let go of or give away while they are weeding in the garden or deadheading. You can ask them to plant something at the start of a new endeavor and track their own development along with the growth of the plant. If you are working with someone who feels isolated from their community or family, you can ask them to initiate an activity that teaches them about their connection to the earth. You might ask them to reflect upon how the sun, soil, plants, rocks, and air support their life. Where do they see themselves in the natural world? You can also help them cultivate images that support this feeling of interconnectedness. Ask them to imagine that the plant they are working with has a voice and can speak to them. What might this plant want to tell them? What might they want to say to the plant? Stephen and Rachael Kaplan explain that nature supports our use of indirect attention (Kaplan and Kaplan 1989). Nature cultivates our capacity for reflection and for planning the course of our life. It allows us to see the big picture we often miss when engaged in our daily-directed activities. From this vantage point, we are able to access wisdom regarding our life journey.

Contributed by J. S. Rice (Figure 1.6)

Figure 1.6 A simple seed-planting activity can offer many therapeutic opportunities.

Exhibit 1.7   Horticultural Therapy Practice: Relationship to Plants

Definition of healing

It is important to recognize that our definition of healing influences our work with others. If we define healing as restoring someone to their original state of health, we may find that we often feel like we are failures or that our work is having little impact. This may particularly be the case if we are working with people with disabilities or those who have disorders brought on by the aging process. An alternative definition of healing is restoring one to the experience of wholeness and connection to the cycle of life. If our model of health implies physical perfection, we needlessly suffer. Our work in the garden and experiences in nature show us that things are rarely perfect in a "national park picture" way (Lopez 1989). If we enter into the wilderness, we find trees that were struck down by lightning valiantly growing toward the light. In our surroundings, we observe the breakdown and decay that supports new cycles of growth, as well as a teeming clutter that calls into question our heroic attempts to render life neat and orderly. If our model of wholeness connects us to this ongoing, vibrant, and sometimes chaotic natural impulse toward life that includes birth, life, death, and rebirth, we will help foster healing in those with whom we work. What calls out for healing are the beliefs we hold that cordon off disease, disabilities, aging, and death from our understanding of life, nature, and community (Berry 1983, Winkler 2003).

How cultivating plants teaches us about healing

When we face serious life-threatening or disabling illnesses or diseases, we often become immobilized with fear and depression and withdraw from life. Our sense of self is affected emotionally, physically, mentally, and spiritually (Sacks 1984). Cultivating a relationship with plants catalyzes movement in each of these aspects of our being (Rice 2001). When people feel diminished in some way, they may feel they have lost the ability to offer something meaningful to the lives of others. Caring for plants provides physical confirmation of our ability to care for life (Rice 1993). Physically caring for plants also provides an outlet for pent-up emotions. In essence, it is possible to give our emotions away to plants (Eagle Star Blanket 2004). Physical activity also stimulates the production of endorphins that elevate our mood. Plants also provide us with an opportunity for reflection (Kaplan and Kaplan 1989). As we work with them, we deepen our understanding of the life cycle. Plants teach us that every living being goes through periods of new beginnings, growth, emptiness, death, and rebirth. As we cultivate plants, they reciprocate by cultivating within us a deeper connection to the seasons of our life. In so doing, they inspire identification with the spirit of life that resides in all living things (Gibson 1989, Rice 2001).

Contributed by J. S. Rice

are needed in the areas of plant science, soils, indoor and outdoor gardening techniques, ornamental and food crops, enabling garden techniques, pest control, and organic growing methods. Therapists may also need to know how to apply information about universal design, greenhouse or nursery management, and therapeutic landscape principles (Marcus and Sachs 2014). A wide repertoire of knowledge and gardening experience is necessary to select the most effective treatment methods and activities in horticultural therapy practice (Starling et al., 2014).

Summary

The field of horticultural therapy may be compared with an iceberg. When a client engages in a horticulture activity, the actions, and sometimes responses, are visible. We can see the garden flourish. We may be able to see attention, smiles, or calm expressions on the faces of participants. We may hear positive social interaction or expressions of joy (Figure 1.7). Hidden beneath the surface are two very large aspects of horticultural therapy—the power of the relationships between the person, plant, and therapist, and the process and planning that accompany the successful horticultural therapy session. On the surface it looks simple; underneath are complex reasoning and processes as well as relationships that make this modality a compelling tool for working with people. As the horticultural therapist plans and sets up a gardening activity based on the needs and skills of participants, a client or patient experiences the planting of a seed, digging in the soil, or tasting a ripe strawberry. A catalyst for positive change is thus set in motion (Figure 1.8).

These techniques and processes that underlie the visible parts of horticultural therapy are described in later chapters and are intended to be applicable (with some adaptation) across the full range of programs. The horticultural therapy profession has evolved into one that offers varied treatment methods and serves a wide array of client groups. Programs

Figure 1.7 Relationships and session planning enhance therapeutic encounters in the garden.

Figure 1.8 Skills of the therapist are integral for the activity to become a springboard for human growth. Photo by Rebecca Haller.

are found in many types of health-care, human services, and community settings.

To provide a framework for succeeding chapters, this chapter has

- Discussed some influences on the profession of horticultural therapy
- Defined horticultural therapy
- Noted the importance of understanding the people served by horticultural therapy programs
- Described some important characteristics about the inherent value of using a horticulture medium for therapy and human development

References

American Psychiatric Association. eds. 2000. *Diagnostic and Statistical Manual of Mental Disorders* DSM-IV-TR (text revision). Washington, DC: American Psychiatric Association.

Baker, Howard S. and Margaret N. Baker. 1987. Heinz Kohut's self psychology: An overview. *American Journal of Psychiatry* 144: 1–9.

Berry, Wendell. 1983. *A Place on Earth*. San Francisco, CA: North Point Press.

Bronfenbrenner, Urie. 1979. *The Ecology of Human Development: Experiments by Nature and Design*. Cambridge, MA: Harvard University Press.

Darwin, Charles. 1998. In Ekman, Paul ed. *The Expression of the Emotions in Man and Animals*. 3rd edn. New York: Oxford University Press.

Diehl, Elizabeth R. Messer. ed. 2007. American Horticultural Therapy Association. http://ahta.org/sites/default/files/Final_HT_Position_Paper_updated_409. pdf (accessed May 20, 2015).

Dorn, Sheri and Diane Relf. 1995. Horticulture: Meeting the needs of special populations. *HortTechnology* 5(2): 94–103.

Eagle Star Blanket. 2004. *Trail of Prediction: Earth Passage*. Conifer, CO: Eagle Dreams.

Einstein, Albert. 1995. The goal of human existence. In *Out of My Later Years: The Scientist, Philosopher, and Man Portrayed Through His Own Words*. New York: Carol Publishing Group.

Gibson, Roberta. 1989. *Home Is the Heart*. Rochester, VT: Bear & Co.

Hagedorn, Rosemary. 1995. *Occupational Therapy: Perspectives and Processes*. Crawley, UK: Churchill Livingstone.

Haller, Rebecca L. 1998. Vocational, social, and therapeutic programs in horticulture. In Simson, Sharon P. and Martha C. Straus, eds. *Horticulture As Therapy: Principles and Practice*. Binghamton, NY: The Food Products Press.

Haller, Rebecca L. 2003. Advancing the practice of horticultural therapy. Unpublished presentation at AHTA Annual Conference.

Kaplan, Stephen. 1978. Attention and fascination: The search for cognitive clarity. In Kaplan, Stephen and Rachel Kaplan, eds. *Humanscapes: Environments for People*. North Scituate, MA: Duxbury Press.

Kaplan, Rachel and Stephen Kaplan. 1989. *The Experience of Nature: A Psychological Perspective*. New York: Cambridge University Press.

Kohut, Heinz, Arnold Goldberg, and Paul E. Stepansky. 1984. *How Does Analysis Cure?* Chicago, IL: University of Chicago Press.

Lewis, Charles A. 1996. *Green Nature/Human Nature: The Meaning of Plants in Our Lives*. Urbana, IL: University of Illinois Press.

Lopez, Barry. 1989. *Crossing Open Ground*. New York: Vintage Books.

MacLean, Paul D. 1973. *A Triune Concept of Brain and Behavior*. Toronto: University of Toronto Press.

Marcus, Clare Cooper and Naomi A. Sachs. 2014. *Therapeutic Landscapes: An Evidence-Based Approach to Designing Healing Gardens and Restorative Outdoor Spaces*. Hoboken, NJ: Wiley.

Matsuo, Eisuke. 1992. What we may learn through horticultural activity. In Relf, Diane, ed. *The Role of Horticulture in Human Well-Being and Social Development: A National Symposium*. Portland, OR: Timber Press.

Matsuo, Eisuke. 1999. What is "horticultural wellbeing"—in relation to "horticultural therapy"? In Burchett, Margaret D., Jane Tarran, and Ronald Wood, eds.

Towards a New Millennium in People-Plant Relationships. Sydney: University of Technology, Sydney Printing Services.

Mattson, Richard H. 1982. A graphic definition of the horticultural therapy model. In Mattson, Richard H. and Joan Shoemaker, eds. *Defining Horticulture As a Therapeutic Modality*. Manhattan, KS: Kansas State University.

Mobily, Kenneth E. and Richard D. MacNeil. 2002. *Therapeutic Recreation and the Nature of Disabilities*. State College, PA: Venture Publishing.

Muir, John. 1969. *How to Keep Your Volkswagen Alive: A Manual of Step-by-Step procedures for the Complete Idiot*. Emeryville, CA: Avalon Travel Publications.

Myerhoff, Barbara. 1978. *Number Our Days*. New York: Simon & Schuster.

National Wellness Institute. *About Wellness*. http://www.nationalwellness.org (accessed June 30, 2015).

Neuberger, Konrad R. 2008. Some therapeutic aspects of gardening in psychiatry. *Acta Horticuturae* 790: 109–13.

Olszowy, Damon R. 1978. *Horticulture for the Disabled and Disadvantaged*. Springfield, IL: Charles C. Thomas.

Paul, Stanley and David P. Orchanian. 2003. *Pocket Guide to Assessment in Occupational Therapy*. Clifton Park, NY: Delmar Learning.

Rice, Jay S. 1993. Self-development and horticultural therapy in a jail setting Unpublished doctoral dissertation. San Francisco School of Psychology, San Francisco.

Rice, Jay S. 2001. A question of balance: Human–plant relations in the soul's journey. Unpublished presentation at AHTA Annual Conference.

Rice, Jay S. 2012. The neurobiology of people–plant relationships: An evolutionary brain inquiry. *Acta Horticuturae* 954: 24–28.

Sacks, Oliver. 1984. *A Leg to Stand on*. New York: Touchstone Books.

Sempik, Joe, Jo Aldridge, and Saul Becker. 2003. *Social and Therapeutic Horticulture: Evidence and Messages from Research*. Leicestershire, UK: Media Services, Loughborough University.

Simson, Sharon P. and Martha C. Straus. eds. 1998. *Horticulture As Therapy: Principles and Practice*. Binghamton, NY: The Food Products Press.

Springer, Lauren. 1994. *The Undaunted Garden: Planting for Weather-Resilient Beauty*. Golden, CO: Fulcrum Publishing.

Starling, Leigh Anne, Tina Marie Waliczek, Rebecca Haller, et al. 2014. Job task analysis survey for the horticultural therapy profession. *HortTechnology* 24(6): 647.

Thrive. 2015. What is social and therapeutic horticulture? http://www.thrive. org.uk/what-is-social-and-therapeutic-horticulture.aspx (accessed May 20, 2015).

Ulrich, Roger S. 1993. Biophilia, biophobia, and natural landscapes. In Kellert, S.R. and E.O. Wilson, eds. *The Biophilia Hypothesis.*

Winkler, Gershon. 2003. *Magic of the Ordinary: Recovering the Shamanic in Judaism*. Berkeley, CA: North Atlantic Books.

Winkler, Gershon. 2005. On counting Omer. *The Walking Stick Newsletter* May.

chapter two

Goals and treatment planning
The process

Rebecca L. Haller

Contents

Introduction

This chapter describes the process of treatment planning used in horticultural therapy programs. The focus is on treatment planning for individuals, a useful approach even when programs are delivered in a group setting. The aims of treatment in horticultural therapy, as in other therapies, are to help each client advance functioning in one or more areas and improve quality of life. In order to measure outcomes, therapists create plans and document achievements for each individual in the group (Figure 2.1).

Once someone is accepted or referred for horticultural therapy services, the treatment process begins. Prior to this point, a determination is made regarding the appropriateness of horticultural therapy programming for that person. Depending on the setting and systems in use, this decision is often made by a case manager, whose role typically includes referring clients for services, gathering information, coordinating and monitoring treatment and outcomes, and ending treatment or discharge.

Treatment teams

Throughout the process, the horticultural therapist may function as part of a treatment team that includes representatives from various disciplines who are involved in the care or treatment of the person being served. The

Figure 2.1 Students in training plan for an upcoming horticultural therapy session. Photo by Rebecca Haller, Horticultural Therapy Institute.   ·

interdisciplinary team (IDT) is used in many health-care, vocational, and educational programs. The team's makeup varies with the setting and individual, but generally includes key caregivers, therapists, family members, and the client himself or herself. Table 2.1 lists examples of additional IDT members. The IDT's responsibilities include the implementation of all phases of the treatment process as well as periodic meetings—all with the aim to maximize the outcome for the client. By using a team approach, caregivers are encouraged to see the client as a whole person. He or she is known as a unique and complex individual with physical, social, psychological, cultural, and spiritual aspects. Increasingly in health care and human services, the client is central to the process and chooses care, goals, and treatment options. A team meeting that includes the client offers the therapist an opportunity to build a relationship with that client and to engage him or her in the therapy process. When the horticultural therapist is not part of the IDT, or when a team does not exist, it is important for the horticultural therapist to establish and use some form of regular communication with other parties or caregivers associated with the client. Access to case records and/or key information about the client is necessary for effective and safe programming.

Table 2.1 Horticultural therapy practice: Interdisciplinary team members

Setting or program type	Possible team members (in addition to the client, family, and horticultural therapist)
Public schools	Teachers, nurse, speech therapist, occupational therapist, counselor, social worker
Vocational programs for people with developmental disabilities	Vocational rehabilitation specialist, psychologist, speech therapist, physical therapist, occupational therapist, nurse
Elder care setting	Physician/nurse, social worker, occupational therapist, activities professional
Mental health	Psychiatrist or counselor, social worker, therapeutic recreation specialist
Physical rehabilitation	Physician, physical therapist, occupational therapist, therapeutic recreation specialist, vocational rehabilitation specialist, speech therapist

Process of treatment planning

The team creates plans that are based on assessments of the client's functioning and the expected outcomes or achievements to be made. These may be called individual education plans (IEPs), individual program plans (IPPs), individual treatment plans (ITPs), or other terms, depending on the setting. (Further information about the components of an IEP may be found in Chapter 6, Exhibit 6.2.) Plans are written using terminology that is behavioral or observable in order to clearly define the objectives and actions to be taken as well as to make it possible to measure the results. Plans must reflect the goals of the individuals being served (Figure 2.2).

Figure 2.2 Careful consideration of treatment objectives is necessary for effective programming. Photo by Horticultural Therapy Institute.

Table 2.2 Horticultural therapy practice: The treatment process

Phase	Actions
Assessment	Information gathering and analysis—knowing the client
Goal identification	Prioritization of problems, needs, and aims
Action plan (Intervention plan)	Creation of objectives, actions, and means for measurement
Intervention	Putting the plan into action, horticultural therapy activities, therapist–client–plant interactions
Documentation	Recording of actions (occurs alongside the intervention phase)
Revision	Ongoing review of results and modification of plan
Termination	Discontinuation of treatment, reporting results

As in other disciplines, the treatment process in horticultural therapy follows a logical sequence from assessment to termination as outlined in Table 2.2.

Beginning with the *assessment*, or information-gathering phase, a foundation is laid for all interventions to come. This phase allows the horticultural therapist to build a solid program for the client based on needs, desires, and abilities. Information may come from standardized assessment tools, interviews, observation, and/or case records. Through a process of careful review and discussion with the client, a problem or set of issues emerges. Particularly when many issues are present, choosing the one to be addressed requires care, skill, and thoughtfulness on the part of the horticultural therapist and other team members. It may be helpful to create lists and narrow the focus through a process of elimination (Buettner and Martin 1995). Begin by listing needs and concerns that have been expressed. Also list the strengths and interests of the client. With the latter in mind, begin to eliminate those concerns on the first list that are

- Low priorities for the client (or even unacknowledged by him or her as "problems")
- Relatively irrelevant to the client's quality of life
- Impractical to address in the horticultural therapy setting or program
- Not likely to be successfully improved

This process, along with communication with the client and caregivers, usually narrows the list to a few priorities. Generally, it is recommended that clients work on no more than three objectives at a time. The more focused the treatment, the more likely success will be met with measurable outcomes. A treatment focus does not eliminate the need to constantly be aware of the person as a whole human being. He or she is not just the "problem" being addressed. In all interactions and horticultural

Figure 2.3 Horticultural therapists regard each person individually and holistically. Photo by Anna Terceira.

therapy sessions, the therapist is mindful that each person and situation is unique (Figure 2.3).

The client's aims, concerns, and dreams must be considered. Even if a dream seems unrealistic at first, look for intermediate steps that move toward that aspiration. It may be, in fact, possible to reach a dream or modify it to be achievable. Furthermore, a client whose wishes are respected will be more motivated to work on the steps involved in therapy. The concept of "culture change" in older adult services includes a focus on programming that is both person-centered and self-directed. (See more about "culture change" as an important element in choosing horticultural therapy activities and techniques in Chapter 3.)

Once the needs and concerns have been prioritized, *goals* to address them should be clearly stated. Goals are generally long term, and articulate something to be achieved through several steps or a series of efforts. (Exhibit 2.1 shows examples of goals.)

The next step in the process of treatment is an *action plan* that includes detailed objectives or short-term goals, actions to be taken, and

Exhibit 2.1   Horticultural Therapy Practice: Long-Range Goals
- Obtain independent employment in the community
- Garden independently at home following a spinal cord injury
- Maintain friendships within an assisted-living setting
- Manage anger effectively in the school playground

plans for how to measure the outcomes. The action plan allows all parties involved to know what specific behaviors are desired, what actions and interactions will take place to facilitate their accomplishment, and how the progress or results will be recorded. The therapist and treatment team use a problem-solving approach to formulate ways to alleviate the issues facing the client. The following list defines the components of the action plan:

- Objective: A precisely worded statement of what is to be achieved within a short time frame
- Methods: Who will do what, how, under what circumstances, where, when, and how often; includes what the therapist will do and the strategies for intervention
- Criteria: Quantifies achievement of objective, for example, how much, how many, how often, how well (the criteria may be included in the objective itself)
- Documentation: Outlines who will measure, what will be measured, how often, and where, or how it will be recorded

As a key part of the action plan, the *objectives* are clearly written, detailed descriptions of what is to be achieved within a short specific time frame. They are written using behavioral terminology so that progress can be measured, and include specific criteria for achievement. Objectives also include information about what, by whom, and how the actions will be carried out, referred to as "methods."

What makes a good objective? An essential characteristic is that the objective logically *leads to long-range goal attainment*. The team, including the client, must be thoughtful about its selection. Similarly, it must be *achievable and realistic* under the circumstances. It must be *precisely worded* to avoid confusion and disagreement. It must be *measurable* so that outcomes can be documented accurately. And, the *client must agree* to participate. (Examples of objectives, based on the long-range goals from Exhibit 2.1, are shown in Table 2.3 and Figure 2.4)

In order for objectives to be clear to all parties involved and to facilitate the recording of outcomes, they should be written following the

Table 2.3 Horticultural therapy practice: Objectives

Long-range goals	Example objectives
Obtain independent employment in the community	Increase speed (while maintaining accuracy) on specified greenhouse tasks by 20% of current baseline rate for two consecutive weeks
Garden independently at home following a spinal cord injury	Use extended-handle gardening tools to plant bedding plants in prepared raised beds without assistance for three sessions
Maintain friendships within an assisted-living setting	Initiate a casual conversation with at least one "garden club" member during each session for one month
Manage anger effectively in the school playground	Control anger in the playground at recess by independently walking to and sitting quietly in the garden, away from situations and people that elicit outbursts, 90% of the time for four consecutive weeks.

Figure 2.4 Participants work to increase speed on a greenhouse task for the purpose of obtaining employment. Photo by Rebecca Haller.

guidelines in Exhibit 2.2. (Further directions for writing effective goals are discussed in Chapter 6 as SMART goals.)

The next step in the treatment process is *intervention*—carrying out the plan. This is when the horticultural therapy takes place—when the horticultural therapist, client, and plants interact with the purpose of reaching stated objectives. For most horticultural therapists and clients, this is the focus—the part of the process that is the most rewarding and enjoyable. Program and activity planning may take place at this stage, always based on treatment plans and objectives for the individuals involved.

**Exhibit 2.2   Horticultural Therapy Practice:
Writing Measurable Objectives**

For an objective to be functional, it should

1. State the *desired behavior* or performance. Indicate what the client will do, what behaviors are expected. Include an action verb. Focus on one behavior per objective.
2. Identify *conditions* under which the behavior is to be performed. These could include where, or with what support, the action is performed.
3. Include *criteria* or standards to measure performance. Identify the criteria for acceptable performance such as for how long, with what accuracy, how much, how well, and so on. (Criteria may be included in the objective statement itself or listed separately on the action plan.)

The objective should be understandable by others, clearly specifying the desired behaviors. It should not include information about what the therapist will do, unless some form of support is included in the conditions. This information is part of an action plan or strategy for intervention, but is not included in the objective itself.

(Chapters 3 through 5 provide more detail about developing horticultural therapy sessions, offering horticultural activities, and working with program participants to motivate and encourage goal attainment.) During the course of engaging in horticultural activities with participants, surprising behaviors may emerge as connections and relationships are built or strengthened. One of the marvels of using horticulture as a therapeutic tool is how a connection with nature can affect someone deeply and profoundly. Therefore, the horticultural therapist must be constantly alert for opportunities to enhance the therapeutic experience for each client served as well as to carry out the action plans as written (Figure 2.5).

Concurrently with intervention, behaviors specified in the action plan are *documented*. Records are kept according to schedule and are used to show progress (or lack of progress) toward stated objectives. What is recorded depends on the objectives in place. The agency or institution where the horticultural therapy occurs often dictates which forms or procedures are used. The important point is that objectives and documentation are aligned. In other words, the data recorded must measure progress toward the objective. In some instances, additional program information may be regularly noted, such as attendance, productivity, or notes about

Figure 2.5 Horticulture can be an effective tool for building relationships. Photo by Pam Catlin.

the process and activity. Documentation should reflect observable behaviors rather than subjective impressions or inferences. It should be kept confidential and only discussed with parties that are authorized to access that information such as the treatment team and the client. (Documentation is discussed further in Chapter 6.)

Throughout the intervention stage of the treatment process, the therapist must monitor the progress of the client. Using good documentation as a basis, the therapist notes achievements and/or lack of advancement and evaluates the objectives and action plan. In the *revision* phase, it may be necessary to modify target dates, criteria, methods, or even the objectives themselves to ensure growth and development. The client may progress faster or slower than anticipated, may have reached a ceiling of achievement, or may have experienced other influences that affect advancement. If objectives have been met, the client should move to the next step, as appropriate. For example, other steps toward the long-range goal in Table 2.3, "obtain independent employment in the community," might be to "begin work on own initiative" or to "get to the worksite independently." The agency may specify procedures regarding revision of objectives or action plans. In some cases, the therapist involved makes simple adjustments routinely. In others, the IDT must meet to consider the changes or a case manager must approve them.

The final phase of the treatment process is *termination* or *discharge*. Horticultural therapy services may end for various reasons, including discharge of the client from the care setting, ending of a seasonal program, client progress (or regress) to a new program, goal achievement, or desire of the client. Again, procedures for ending horticultural therapy participation follow those of the employing agency. Generally, the therapist helps to prepare the client for the transition, reviews documentation, and may write a summary report that includes objectives, achievements, and recommendations for other services, as needed.

Summary

Horticultural therapy is practiced in diverse settings, under many different circumstances. This chapter has outlined a process for providing horticultural therapy services that can produce measurable outcomes for program participants. In some settings, therapists need to modify this process in order to fit in with existing methods. In summary, horticultural therapy programming should be individualized and based on prioritized needs or concerns, be goal driven, be articulated clearly, and documented with methods that show measurable results—all within the framework of the organization with which it is associated. This will allow the horticultural therapist to serve clients well and to communicate results in a coherent and professional manner.

Bibliography

Austin, David R. 1991. *Therapeutic Recreation: Processes and Techniques.* Champaign, IL: Sagamore Publishing.

Borcherding, Sherry. 2000. *Documentation Manual for Writing Soap Notes in Occupational Therapy.* Thorofare, NJ: SLACK.

Buettner, Linda and Shelley L. Martin. 1995. *Therapeutic Recreation in the Nursing Home.* State College, PA: Venture Publishing.

Davis, William B., Katie E. Gfeller, and Michael H. Thaut. 1992. *An Introduction to Music Therapy: Theory and Practice.* Dubuque: Wm. C. Brown Publishers.

Hagedorn, Rosemary. 1995. *Occupational Therapy: Perspectives and Processes.* New York: Churchill Livingstone.

Hagedorn, Rosemary. 2000. *Tools for Practice in Occupational Therapy: A Structured Approach to Core Skills and Processes.* New York: Churchill Livingstone.

Hogberg, Penny and Mary Johnson. 1994. *Reference Manual for Writing Rehabilitation Therapy Treatment Plans.* State College, PA: Venture Publishing.

Mager, Robert F. 1997. *Preparing Instructional Objectives.* Atlanta, GA: The Center for Effective Performance.

Ozer, Mark, Otto D. Payton, and Craig E. Nelson. 2000. *Treatment Planning for Rehabilitation: A Patient-Centered Approach.* New York: McGraw-Hill.

Stumbo, Norma J. and Carol Ann Peterson. 2004. *Therapeutic Recreation Program Design: Principles and Procedures.* 4th edn. San Francisco, CA: Pearson Education.

Zandstra, Patricia J. 1988. A systematic approach to horticultural therapy. *Journal of Therapeutic Horticulture* III: 15–24.

chapter three

Activity planning
Developing horticultural therapy activities and tasks

Pamela A. Catlin

Contents

Introduction

Horticultural therapy facilitates connection between people and plants—between people and nature. This connection is the essence of how and why gardening activities are such powerful catalysts for positive human development. The garden provides people with a multitude of opportunities for seeing themselves, and the world, from a new perspective. Its associated activities can be used for consciously enhancing spiritual connections and growth. As activities are chosen, it is imperative that therapists look first to the garden to find those natural activities that enhance this relationship (Figure 3.1).

With this in mind, this chapter explores many of the variables that go into developing effective horticultural therapy activities, including

- Goals addressed in the horticultural therapy program
- Selecting activities or tasks to meet therapeutic goals

Figure 3.1 Gardening is a natural and powerful change agent.

- Additional considerations for activity selection
- Planning for seasonality
- Resources for session ideas

Horticulture as a vehicle to meet wide-ranging goals

Because of its broad-ranging appeal and varying levels of skills required, horticulture lends itself to serve the needs of many diverse populations and ability levels. Some of the primary goal areas addressed in successful horticultural therapy programs are

- Physical
- Cognitive
- Sensory stimulation
- Emotional
- Interpersonal/social
- Community integration

Physical

Some of the physical benefits derived from a horticultural therapy program are obvious to the casual observer. Being out in a beautiful garden environment in the fresh air is most likely what comes to mind. However, many other physical goals and objectives can be met through the horticultural therapy program. Examples are maintenance or improvement of fine motor skills, gross motor skills, standing/balance and endurance, mobility, range of motion, nutrition, and strength (Figure 3.2).

Figure 3.2 Watering the garden offers a purposeful and relaxing activity to build endurance and strength. Photo by Pam Catlin.

Often, therapists will find participants willing to pursue physical goals with greater ease in the garden than in the therapy room. The familiarity that a garden can bring, the sensory stimulation that is part of being surrounded by and working with plants, the distraction and focus of working with plants, and the reward of the beauty and harvest—all can play a part in this process.

Cognitive

Cognitive goals address areas such as

- Speech and vocabulary
- Memory
- Learning new skills
- Sequencing
- Following directions
- Problem-solving
- Attention to task
- General brain functioning

Incorporating information about the plants being used, such as their names and where they come from geographically, are ways to enhance a program. Orientation to time and place can be facilitated through the horticultural therapy program by incorporating special holidays and seasons as well as "checking in" on the progress of planting projects to give participants a gauge for time. Simple tasks such as writing one's name, the date, and plant name on a label can accomplish several cognitive goals in one aspect of an activity. Choosing tasks that fit the neurodevelopmental needs of a client provides opportunities for foundational personal growth (Rice 2012) (Figure 3.3).

Sensory stimulation

The horticultural therapy program can be an excellent avenue for providing sensory stimulation. Often a required component of activity programs, sensory stimulation can be met through the use of fragrant flowers and herbs, plants with interesting textures and tastes, and specific color combinations. Sometimes overlooked, the auditory senses can be awakened by designing the garden to have trees and grasses that create subtle sounds or by using a water meter indoors that makes audible clicking sounds when plants are wet. The goal of sensory stimulation can be tied to both cognitive and emotional goals when thinking of activities to be used. For example, a time for sharing the fragrance and taste of various herbs could stimulate reminiscing about a childhood memory of

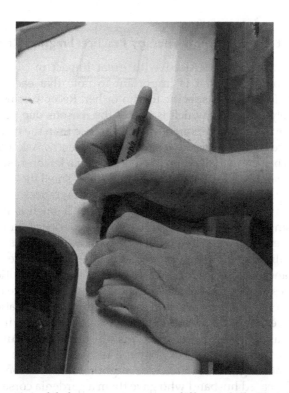

Figure 3.3 Creating a label encompasses many different skills. Photo by Christine Capra.

collecting herbs for one's grandmother. More detail about responses to fragrance is shown in Exhibit 3.1 (Figure 3.4).

Emotional

The horticultural therapy program can help create a stronger sense of self-esteem for participants. In situations where, all too often, choices and responsibilities are taken away, the horticultural therapy program can provide a safe way to incorporate those functions back into people's lives. In settings where individuals are most often in the care-receiver role, the horticultural therapy program can provide an outlet for the participant to give care to the plants and to other people as well. Esteem building can take place through the unlimited opportunities for creative expression offered in a diverse horticultural therapy setting. Horticulture activities, more than many other hobbies, are often appropriate for any age and can easily be adapted to any ability level. Well-planned and implemented age-appropriate horticultural activities result in enhanced self-esteem (Figure 3.5).

Exhibit 3.1   Horticultural Therapy Practice: Fragrant Responses

Fragrance can be a tool used with great impact in horticultural therapy programs. But it is important to note that each person's reaction to fragrance is specific to him or her. Receptors in the nose are uniquely arranged, which is one of the reasons one person can detect more of some components in a fragrance than the next person and why we each experience fragrances differently. Ask a group to describe what they smell when presented with a lemon-rose scented geranium. Some will smell the rose, others will smell the lemon predominantly, and still others will detect other fragrances.

Cultural background can impact whether a smell is perceived as a pleasing or unpleasant odor. Cultural experience with certain foods, for example, affect whether one enjoys or "turns up their nose" at culinary herbs such as rosemary, oregano, or sage.

Fragrance also has the power to evoke powerful emotions tied to memories of specific times, places, and people. While the therapist usually intends happy, pleasant, and comforting associations, the emotional response can be either positive or negative. It is important to be prepared when the associations are unpleasant, uncomfortable, or sad. A gardenia flower might be associated with the pleasant memories of a high school dance, or might remind someone of their recently deceased husband who gave them a gardenia corsage every Mother's Day. When this happens, the horticultural therapist should respond compassionately and allow the person to express his or her feelings about the memory.

Sometimes, the association is linked to an experience. Preference for a particular plant or flower relates to the individual's perception of the situation. A plant that illustrates this rather dramatically is *Plectranthus*, or "Vicks Plant." It generally induces positive feelings for those who remember being comforted by their mother with a menthol rub while they were sick. Others associate the smell with feeling miserable and don't like the plant at all. It is helpful to have a variety of choices so that each participant can choose a plant or flower with positive associations.

Contributed by K. Kennedy

Figure 3.4 Scented geraniums in a therapy greenhouse provide fragrance.

Figure 3.5 Caring for plants gives individuals a sense of purpose. Photo by Pam Catlin.

The horticultural therapist can structure a program to provide emotional benefits in the areas of anger or aggression management. Physical activities can utilize the energy found in these emotions and channel it into productive directions. Examples of successful activities in this area are digging to prepare the garden, weeding, raking leaves, mixing soil, or washing pots.

**Exhibit 3.2   Horticultural Therapy Practice:
Spiritual Connection**

The horticultural therapy program is an ideal setting for individuals to explore their own spiritual connection, and this can be profoundly transformational. Spirituality, not to be confused with organized religion, has to do with feeling a sense of connection with one's surroundings, with others, and with the natural cycles of life. For as long as there is recorded history, the spiritual connection between plants and people has appeared in art of all forms. In the fast-paced, man-made world or the typical institutional world, life can seem out of balance. There is an order in nature. The symbiotic relationship between the birds, butterflies, bees, and flowers, or the perfection in a flower, can help an individual to regain a sense of being at peace and renew a sense of awe and interest in the world of which they are a part.

The emotional benefit of an improved outlook on life may be fostered through ongoing horticultural activities where individuals are encouraged to look for signs of weekly, if not daily, growth and change. The cycle of propagating garden plants from seed is an example of this process. One's focus can evolve—starting with the needs and changes of the seedlings, to planting in the garden through harvesting. Such projects can provide those involved with a sense of wonder and purpose in life. The garden setting also enhances a spiritual connection through a sense of connectedness to nature and a witnessing of the cycles of life (see Exhibit 3.2). In the wellness arena, the emotionally rejuvenating link that nature provides for most people can be a stabilizing force in a rushed, stressful work environment (Figure 3.6).

Interpersonal/social

The horticultural therapy program can provide social benefits in a number of ways. The simplest is the socialization that naturally occurs when participants are brought together for a group session. There are many anecdotal reports of individuals with histories of being withdrawn who begin speaking and/or participating in a garden task or activity when shown a blooming plant. On more subtle levels, socialization can take place through such things as

- Enhancing family visits through sharing garden happenings
- Creating projects in a session

Figure 3.6 Preparing the garden for spring growth is an activity that brings a nature connection as an antidote to a stressful, rushed lifestyle. Photo by Rebecca Haller.

- Conversing with other people while weeding
- Sharing with others the bounty of the garden
- Cooperating with others in gardening tasks

Another facet of social benefits is recreation or leisure skills. Every person, regardless of age or ability, needs a recreational outlet. Gardening, whether a new skill or a long-time favorite hobby, is an excellent leisure-time activity. It helps motivate people to move out of a sedentary lifestyle. Gardening can be done by oneself or cooperatively with others, and it helps to provide a healthier diet as well as add to the esthetics of a home or facility. Working with plants as a leisure skill is a common link with many other "nondisabled" individuals, reducing the stigmas that many people with disabilities might encounter in their personal leisure lives.

Community integration

Because gardening (indoor and outdoor) is one of the top leisure activities of the general public, many opportunities exist for horticultural therapy participants to integrate into the community through horticulture in a recreational sense. These include tours of local horticultural facilities, involvement with local garden clubs, or participation in county fairs and community gardens.

Horticultural therapy may be able to provide job opportunities within the community because green industry employers (nurseries, landscapers, etc.) need people with basic horticulture skills such as watering, planting, and basic garden maintenance. For example, many programs serving youth at risk, individuals with intellectual and/or developmental differences or traumatic brain injuries, and those who are incarcerated provide on-the-job training, working directly with clients in the job setting.

Ideas and activities that may be used to address some of the overall goals discussed in this section can be found in Appendices I and IV.

Activity or task selection to meet treatment objectives

In order to select horticulture activities that will be effective and advance each individual closer to his or her desired outcomes, take into account the following:

- The *activity is a tool* for treatment.
- The *type of program* influences the selection of activities or tasks.
- *Treatment issues/goals* are paramount considerations for choosing what to do.
- Participants each have *varied backgrounds*, skills, and interests.
- A *situational assessment* may yield information to drive future session content.
- *Program continuity* helps connect sessions for enhanced outcomes.

Beyond activity

Horticultural therapy is often activity based, yet it is of vital importance to remember that the plants and garden are only tools in the therapeutic process, not the end goal. Often, a session developed around sensory stimulation and a greater awareness of one's surroundings can be more effective than a gardening project with many steps that results in something to carry out the door. There may be times in specific settings when an activity is used for diversion, but far more often the horticultural therapy activity is devised to assist individuals to achieve specified outcomes.

When therapists ask horticultural therapy participants to expend energy in therapeutic groups, they need to have clear ideas about what makes an activity both appealing and useful (Borg and Bruce 1991). The activity should be viewed as therapeutic by the therapist using it and should be tied to an acceptable therapeutic theory and base. When therapists are fully engaged with the activity, they communicate that the activity is worthwhile and can lead to goal attainment. Desirable characteristics of an activity are that it

- Is purposeful
- Is interesting
- Offers a chance to take responsibility
- Necessitates an investment of energy
- Can be task analyzed (as described in Chapter 4)
- Is balanced between demands and patients' abilities
- Has a beginning and an end

Type of program or facility

The type of program, organization, or site plays a major role in determining the selection of an activity or task. The following is a listing of the many types of facilities where horticultural therapy is used, with examples of the programs typically found in those settings.

Vocational organizations

Vocational programs might have a horticultural therapy program with an emphasis on training people in specific horticultural job skills such as potting, transplanting, watering, fertilizing, and so on. The horticultural setting is also often used to provide individuals with basic job training including the development of social skills. Additionally, there is a trend for these programs to provide opportunities for learning new leisure skills, in which case a more recreational and social approach to activity selection would be necessary. See Exhibit 3.3 for an example of how vocational programs may address the needs of the whole person.

Long-term care

In the long-term care setting, a recreational or wellness program is often utilized. Activities ideally address general goals such as increased socialization, sensory stimulation, and cooperation with others. A therapeutic emphasis can also be implemented by selecting activities that address goals such as improved or maintained motor skills, cognitive skills, or orientation to time and place. This focus can take place either in the group setting or individual sessions. In long-term care, maintenance of skills is usually the goal. There often are, however, individuals who might reside at a care community for a brief period of rehabilitation. In this case, a focus on improvement of skills would be more appropriate. In long-term care settings, most individuals will be at the site for an extended period of time. From a horticultural therapy perspective, this provides

Exhibit 3.3   Horticultural Therapy Practice:
Goal Areas in Vocational Program

While a program may have primary target goals, it is not uncommon for a program to be able to address goals in many different areas over the course of time. This is particularly true of long-term treatment programs. For example, within a day treatment, prevocational program for adults with traumatic brain injury, a garden area is maintained containing plants to be harvested for a fall herb vinegar sale. The following goal areas may be addressed:

- Physical: Balance, coordination, and endurance
- Cognitive: Practicing compensatory strategies for memory, planning and organizational skills, and writing skills
- Emotional: Anger management, impulse control, stress management, sense of accomplishment and success, self-esteem, and nurturing opportunities
- Spiritual: Sense of connection to one's environment, regain life balance
- Interpersonal/social: Communication skills, relating to peers, cooperation, teamwork
- Community integration: Appropriate behavior in a retail environment (purchasing the plants), relating to others who use the garden
- Recreation/leisure: Learn transferable skills to be used at home, interest in a new hobby

Contributed by K. Kennedy

opportunities for activities that carry over from one week to the next, providing a continuity to the program. Propagating cuttings for dish gardens or terrariums or planting in the spring for a fall harvest are just two examples of long-term activities.

Short-term or acute care

In short-term care, the emphasis may be for a patient to build endurance, increase skills, or cope with a medical condition or newly acquired illness. The therapist may take a psycho-educational approach, providing training in these and other areas. Ideally, activities need to fit into the time frame of a patient's stay. When incorporating long-range projects such as planting the outdoor garden or forcing bulbs, the therapist would need to help the participants understand that they are part of a bigger picture.

Although they might not be available for harvest or full bloom, it should be communicated that they are an important part of the process and play a vital role in it. They can also benefit from work done by previous program participants. In these situations, the therapist can remind patients of the goals they have set and explain how those goals can be met by participating in only a segment of a long-range activity.

Psychiatric

A therapeutic focus using plants as metaphors to promote self-care, life skills, or a plan for developing responsibility (thus building self-esteem) could be key in these programs. Activities and tasks need to meet the safety requirements of the facility. This could involve the materials and tools used as well as considerations regarding certain medications (Figure 3.7).

Physical rehabilitation

A therapeutic focus on rebuilding motor skills, self-esteem, spatial relationships, cognitive development, stamina, improved concentration, and use of adaptations is used in this setting. The suggestions found in short-term and acute care could also be useful.

Schools

Study of the environment and group cooperation/education are often the focus in school programs. Activities can be selected that help to meet these goals, at the same time developing and improving social and coping skills. Horticultural therapy programs in high school settings often provide prevocational training for students with disabilities. In these programs, basic horticultural knowledge and work skills are emphasized.

Figure 3.7 A vocational approach to treatment for substance abuse may include greenhouse production and sales.

Corrections

Programs in the field of corrections are typically vocationally based, teaching the job skills required for greenhouse, nursery, or landscape employment. Often large gardens are incorporated into the plan by providing opportunities to raise vegetables to be shared with people in need through food banks or homeless shelters.

Wellness or health promotion

In the wellness arena, a holistic approach is taken whereby horticultural-based activities focus on stress reduction, coping skills, regaining choice, and control. Reintegration with nature can result in a regeneration of energy and self-care.

Community gardens

Horticultural therapy can be incorporated into a community garden by using the garden as a site for any type of program. Many community gardens focus on community development and serve disenfranchised populations as well as people with a passion for gardening. Coordinators may develop specific measurable goals and objectives for participants and create a program based on those. Some areas of focus that would fit well into a community garden setting include

- Activities with an *educational* focus regarding proper nutrition
- A *therapeutic role* using metaphors in the garden for life-skill development
- A *social/emotional* role in adding garden-related recreational activities to the schedule

Treatment issues/goals

Other key factors that influence the selection of horticultural therapy activities or tasks are the individual goals to be addressed. In many settings, individuals are given some form of a needs assessment when they begin treatment (as described in Chapter 2). This identifies each person's strengths and areas of concern, often called *treatment issues*. Examples of some treatment issues that might be derived from such an assessment are needs for

- Increased socialization
- Increased display of appropriate social skills
- Increased orientation to time and place
- Increased ability to follow multiple-step directions
- Increased healthy sense of responsibility and decision-making skills
- Increased cognitive skills

- Decreased anxiety
- Increased sensory stimulation

Most often, a needs assessment will include not only abilities and treatment issues but also information regarding a person's hobbies and gardening interests.

Participants' backgrounds, likes, and dislikes

The horticultural therapist may also provide an assessment designed specifically for the horticultural therapy program. This would help in obtaining a clearer understanding of a person's interest and experience regarding indoor and outdoor gardening. For gardening to have therapeutic value, it must be meaningful to the individual participant. This will occur when the individual's needs are assessed and correctly matched to an appropriate component of horticulture. This assessment also determines if there are potential challenges such as an aversion to getting dirty or allergy issues. It is only natural that a person will participate if he or she is interested in what is being done. Ideally, a horticultural therapist takes into account the background of those being served, finding out the extent to which they have gardened in the past, the types of tasks and activities they enjoy, and those they do not.

When beginning to work with an individual, selecting familiar, enjoyable horticultural activities will help to create a sense of safety and belonging for the participant. After a person has been in the horticultural therapy program for a period of time, it might be beneficial to expand the focus of activities to include new and unfamiliar ideas. Providing the opportunity to learn new skills and new concepts can enhance a client's overall functioning level. In the "person-centered care" model (as described in Exhibit 3.4), understanding a participant's background experiences and knowledge is key to providing purpose and meaning to his or her current life. This background knowledge is also valuable when attempting to understand what is behind certain behaviors. *Culture change* is a term that incorporates "person-centered care" and "person-directed care." Though the term *culture change* is identified with the older adult population, aspects of it can be used universally (see Exhibits 3.4 and 3.5 for perspective and ideas for incorporating these concepts into a horticultural therapy program).

Situational assessment

After the general assessment has been completed, the horticultural therapist may carry out a situational assessment. Formal or informal observations are made of an individual actually performing horticulture

Exhibit 3.4   Horticultural Therapy Practice: Person-Centered Care

Person-centered care is a model of care that focuses on the value of each individual. It involves respecting and honoring the uniqueness of each person and allowing him or her to be involved in decisions that impact his or her life. Traditional care and person-centered care differ in the following ways: disease focused versus person focused; managing behaviors versus acceptance and understanding of behaviors; caregiving versus care partnering; control and losses versus empowerment and abilities programming; and activities versus meaningful occupation. In the *American Journal of Alzheimer's Disease*, Virginia Bell and David Troxel wrote: "The goal of person-centered care is to move the person, even momentarily, from loss to fulfillment, loneliness to connectedness, sadness to cheerfulness, confusion to orientation, worry/anxiety to contentment, frustration to peacefulness, fear to security, paranoia to trust, anger to calm, and embarrassment to confidence."

Exhibit 3.5   Horticultural Therapy Practice: Implementing Person-Centered and Person-Directed Care in the Horticultural Therapy Program

1. Know the people in your care as individuals. Use their life experiences with plants/gardens and other areas as a way to connect.
2. Develop goals with participant involvement. Identify desired outcomes by visiting with each person regarding his or her interests and desires. If they are not able to express this verbally, speak with family members and read social histories to glean the information.
3. Provide meaningful activity such as raking, watering, and planting to those who show a desire for a more purposeful experience.
4. Develop projects that can be given to others such as floral arrangements and dish gardens.
5. Provide choices to participants as often as possible.
6. Empower participants to share their knowledge of plants and nature. Simply creating opportunities for them to communicate during group discussions can do this. On a larger scale, a

participant could demonstrate, in front of a group, skills such as floral arranging or preparing a special vegetable dish to share.

7. Create a safe place, emotionally as well as physically, where participants can have a sense of "fitting in." Providing a comfortable setting, matching tasks to ability levels, and creating opportunities for partnering are all means of meeting this goal.

The above are just a few suggestions for incorporating "culture change" into the horticultural therapy program. Two organizations, the Pioneer Network and Leading Age, are excellent resources for obtaining more information on "culture change."

activities. This type of assessment can provide an array of useful information about how the person approaches and accomplishes each activity. It can be used to document skills and behaviors seen in horticultural therapy that may be different from other therapy environments. In the vocational setting (or any task-oriented setting), this may be used to gather data about an individual's work potential in a particular job setting. For vocational programming, the therapist may create a written work sample form that includes the following categories:

- Task
- Objective
- Supplies needed
- Directions to the evaluator
- Directions to the client
- Areas of focus
- Guidelines for scoring

From the data collected, the therapist can then select activities or tasks as part of a training program that will help the client to attain the necessary skills required for the desired job.

Program continuity for individual development

Continuity in the horticultural therapy program is a building block for successful individual development. When a program goes beyond mere activity and becomes a process where activities or tasks interconnect and build on each other, there is a stronger foundation for a person's growth. Continuity adds that dimension to create an identifiable order rather than just a collection of random events. The medium of horticulture is ideal for providing this connection as the garden progresses naturally through the seasons. Continuity can be achieved through

- Programming on a regular basis, ideally at least once a week
- Selecting horticultural activities that build on each other
- Revisiting past projects to note their changes and to orient to time and place
- Using the garden's natural seasonal progression
- Keeping in mind that the activity is part of something larger
- Maintaining consistent workspace and storage areas

Additional considerations for activity planning

In addition to the aforementioned considerations, planners should also take into account the *resources* available at each horticultural therapy site and the number and type of *staff* to be present. Choices that *minimize risk* to all who come into contact with program spaces and materials are essential. Therapists must be *flexible* and use *evidence-based practice* in selecting and presenting activities.

Resources and restrictions within a setting

The setting in which the program takes place also plays a role in activity selection. The following are some of the key factors to take into consideration:

- Time frame allowed for the activity
- Budget
- Other conflicting uses of the programming area
- Availability of water
- Sizes of outdoor and indoor gardening areas
- Hardiness of the growing zone
- Materials, tools, and furnishings available
- Number of participants
- Average length of stay at site for program participants (e.g., a typical stay might be two weeks, or less in short-term psychiatric facilities or women's shelters)

Staffing

The number of staff and/or volunteers available to assist with a session helps to determine what activities are selected. Activities that involve moving through the garden should ideally have a staffing ratio of one-to-one for individuals needing mobility assistance. This ensures safety and a sense of inclusion for all those participating. Those activities requiring extensive materials preparation should be saved for times when volunteer help is available. In activity-based types of programs, tasks that involve a series of steps or procedures are best performed when there is staffing support to distribute materials and provide assistance where needed. This would not be necessary in programs in which it is appropriate to involve the clients in all aspects of the task at hand. Finally, there is the need for plant care between sessions. In programs where the horticultural therapist is only part-time, the level of commitment by participants, staff, or volunteers to water and maintain the plants will determine the types of plant activities selected. A strong commitment would allow for activities that require high maintenance such as propagating seeds indoors or planting a sizable garden. A minimal commitment would direct the therapist to provide activities that can be taken away by the participant on completion of a session and/or to choose low-maintenance plants and garden design, thereby reducing the need for ongoing care by others.

Risk management

In selecting activities, there are criteria to be considered regarding safety in the horticultural therapy program. It is essential to use only nontoxic plants when working with populations that exhibit high levels of confusion, are very young, or are at risk of self-injury. A list of toxic or dangerous plants is often available through local poison control centers and Internet sites. A rule of thumb is "when in doubt, don't grow it." Likewise, one should have an awareness of any toxic properties of materials used in activities or in the garden such as fertilizers or pesticides. The use of sharp utensils and materials is also an area of concern with some populations. Supplies should be used under supervision of staff and tools counted prior to and on completion of a session.

Another area of concern is photosensitivity (sensitivity to the sun). There are a number of pharmaceutical drugs with this side effect that can result in serious sunburns. Some individuals may have intolerance to heat and/or cold due to their medical condition. For example, people with multiple sclerosis may have temperature-regulation issues. Respiratory issues must be taken into consideration when working with fragrant flowers, herbs, or potpourri mixtures. Caution is necessary when using these materials with people with asthma, for example. The horticultural therapy program often incorporates food tasting, so the therapist should be aware of any dietary restrictions and food allergies.

Safety in the garden, greenhouse, and activity area is of utmost importance. Pathways and aisles should be well lit and clear of obstructions so as to be easily traversed by people in wheelchairs or with other mobility issues. The garden and greenhouse areas may also need to be secured in settings such as prisons, dementia care, or some mental health facilities. A careful review of initial assessments and clear communication with key staff and the participants themselves are essential in providing a safe and successful experience.

Flexibility

In addition to the planning suggestions given in this chapter, the therapist should be flexible. An effective therapist has at least one alternative activity planned (Buettner and Martin 1995).

In the field of horticultural therapy, there are many variables. The diversity of participants, weather changes, staffing, and rate of plant growth are just a few of the many factors that can affect plans. Activities can be modified as an effective means of accommodating a group of participants with various ability levels to interact together in the same setting. Adapting the environment, substituting materials, using only parts of activities, or adjusting the type or amount of facilitation provided are a few of the key means of adaptation. More information about adaptations and modifications is found in Chapter 4. Being able to adapt activities is helpful as well as having materials for one or two alternate activities in storage for those times when a complete change is warranted. Some suggested sessions for those times are games such as garden bingo, grooming and feeding plants, pressing flowers, or garden flower arrangements. These, and sessions that involve direct gardening experiences, take a minimum amount of preparation time.

Evidence-based practice

The therapist should utilize information from other practitioners about the activities, plants, and tasks that have been successful for specific persons and program types. Reviewing the literature, newsletters, and social media for this evidence leads to better programming. Budgets are also positively impacted through saving time or materials that might otherwise be spent on less effective activities. More information on evidence-based practice can be found in Chapter 5.

Planning for seasonality and efficient use of horticultural resources

Once the setting and purpose of the program are established and goals have been identified, the therapist may, in some settings, create a calendar of activities for the year. This process helps the therapist to keep the activities in alignment with the goals of the program and the individuals being served and to incorporate horticultural resources in a meaningful way. An activity calendar determines the garden plan and plant selection, produces a congruent program that flows with the cycles of nature, and helps develop an efficient means of purchasing and preparing supplies.

Creating a seasonal calendar

Creating a seasonal calendar is an excellent way to establish order in a horticultural therapy program. The following are considerations in developing a calendar for an activity-focused program:

- List the major holidays that are relevant to the individuals or community being served (see Appendix II for a list of possibilities).
- List horticultural/nature days of note: Arbor Day, Earth Day, the solstices, Horticultural Therapy Week, and so on.
- List celebrations or special events that the organization/facility has in its calendar.
- Research miscellaneous celebrations that might fit in with the population being served. Activity therapists and recreation therapists often subscribe to newsletters that share this information.
- Just as each month has a birthstone associated with it, there is also a flower associated with each month. This could be easily incorporated into the program plan.

Creating a planting calendar

Once the seasonal calendar is developed, an indoor/outdoor garden-planting schedule can be created. Initially, determine what plants will be needed for specific holiday projects or plant sales, such as Mother's Day potted plants, a holiday amaryllis display, Valentine topiary, bedding plants, or cut flowers. Will there be plants needed for miscellaneous plant projects such as terrariums or dish gardens? All of these activities require prior activities that involve propagation of the needed plants. The Mother's Day plants and the Valentine topiary need to be propagated no less than three weeks in advance of their use, and the amaryllises need to be forced in late October for a December display. See Appendix II for a planting schedule of selected holiday products.

The outdoor garden schedule should include plants and planting dates for cool and warm seasons. Identifying when the plants should be transplanted to the garden will determine when to start the seeds indoors. Seed packets as well as books on propagating from seed provide information on seeding times. Once these planting dates have been determined, then activities can be developed for the days remaining open on the calendar. For programs that are not activity focused, crop scheduling may be based on sales, seasons, indoor and outdoor growing space, and the needs of participants.

Coordinating outdoor with indoor horticultural programming

Ideally, program participants should grow most of the items used in the horticultural therapy program. To provide program continuity, the materials used in a winter floral arrangement should be grown and preserved by the participants themselves or, in settings where stays are short, by those who gardened before them. Throughout the growing season, flowers can be pressed or dried. *Helichrysum* (strawflowers), *Gomphrena* (globe amaranth), and other everlasting plants can be harvested, wired, or hung. Herbs can also be harvested and dried. Floral projects provide an opportunity for creative expression as well as for gift giving and continued connection with the garden. Prior to a frost is the time to start bringing the garden indoors (Appendix II includes a table illustrating ways to accomplish this and, in turn, create a rich program). It is wise to treat all cuttings and potted plants with a soap and water solution before bringing indoors to prevent the introduction of insects. Putting the garden "to bed" for the winter is also an important process in which to engage program participants. Incorporating all of the life cycles of the plants and garden in the program offers individuals involved a greater understanding and acceptance of life cycles in general (including their own.)

Idea sources for tasks and activities

An important task for a horticultural therapist is to come up with new ideas for activities. Sources for session ideas abound, including the garden itself as well as the imagination of the program leader (see Exhibit 3.6 on the garden as a source of activities). There are some books available that show activity ideas, and there are many other resources available to the therapist (as shown in Appendix II). An important consideration when looking at a potential activity or gardening task is: "Can the idea be adapted to meet the needs and abilities of the people being served?" (see Chapter 4 for further information about adapting activities). It is essential to remember that the horticulture activity is a tool to assist people toward their goals. Those that require minimal intervention or support by the therapist or volunteers (i.e., those that maximize independence of the individual being served) are the most effective. An activity should be age and ability appropriate and a source of pride for the participant. In a greenhouse or nursery program where items are being produced to sell, the quality of the item is important, and projects can be selected that, again, fit the abilities of those doing the work.

Exhibit 3.6 Horticultural Therapy Practice:
The Garden as a Primary Resource for Activities

Horticultural therapists are planners. Many aspects of professional practice require it. They plan programs as a whole, goals for individuals and groups, treatment techniques, garden design, planting schedules, and so on. One of the most constant planning needs is to come up with activities, session structure, and materials to meet the needs of individuals and groups. Using the garden as a primary resource offers multiple advantages in this ongoing process:

- The garden is genuine, providing real growth, success, and challenges for participants.
- The garden is seasonal, orienting people to place, day, and time of year.
- The garden contains most of the materials needed for purposeful activity, lessening the need to shop for or transport supplies for the session.
- The garden is changing, keeping it fresh and interesting with new delights and things to do and experience.
- Especially true for an outdoor garden, the garden brings outside "visitors" such as birds and butterflies, connecting its growers to a broader world outside the confines of a facility or residence.
- The garden is rewarding. Engaging in all parts of a growing process may foster feelings of stewardship, accomplishment, pride, and belonging.

These are just a few of the benefits to be gained by making use of the garden for session activity ideas.

Contributed by R. L. Haller

Horticulture ideas

In addition to gardening and plant cultivation, a vast array of websites, gardening magazines, and television shows is available, providing the stimulus for interesting and motivational activities. Other excellent resources are programs that produce materials for youth gardening such as the National Gardening Association and the Cooperative Extension Service. Youth gardening information is most often focused on the outdoor garden and is written in a way that is easy to adapt for specific populations served in horticultural therapy. Creating a relationship with managers of greenhouses, community gardens, and horticulture businesses can be an asset for tours and plant materials from which to develop gardening activities. Criteria for selecting horticultural activities are

- Task or activity has therapeutic value and meaning.
- Plants used are safe (nontoxic, no thorns, etc.).
- Plants have the ability to grow in the program environment. Consider plant needs for shade or sun, indoor or outdoor temperature, drought tolerance, and growing space.
- Activity works within budget constraints and time frame. (For example, can the plants be propagated, or will they need to be purchased?)
- Plants are relatively easy to grow and maintain, with a good likelihood of success. Note that in some cases, plants with high maintenance needs may be advantageous because they provide tasks and activities on an ongoing basis.

Floral design and horticulture craft ideas

Websites, blogs, social media, magazines, television programs, and nature craft books are excellent resources for floral design and horticulture craft ideas. Taking time to walk through floral shops, garden centers, and craft stores or shows often pays off with new ideas. When selecting floral or craft activities, be sure the project has therapeutic value and can be adapted for age and ability appropriateness. Both the cost of materials and the possibility to raise the materials in the garden, or to find them in nature, should be considered. Artistic endeavors that utilize materials grown by participants have the potential to be the most rewarding and meaningful (Figure 3.8).

Nature study ideas

Nature or environmental studies can either be the main focus on their own or enhancements for more diverse programming. Gardens are wonderful arenas for observing and connecting with nature. National organizations such as National Arbor Day Foundation and Earth Day Foundation are just two examples of organizations that provide classroom materials and often plants for use in horticultural therapy programs. The National Forest Service is a resource for nature posters, useful in teaching environmental concepts as well as decorating the session area. The environment is another area in which many materials for use with schoolchildren are available and can be adapted to meet the needs of specific populations. Many cities and communities have nature-related organizations such as nature centers or public/private parks that can serve as tour sites or as

Figure 3.8 Working with flowers is commonly included in horticultural therapy programming. (Horticultural Therapy Institute.)

resources for project ideas and materials. Botanic gardens and arboreta serve the general public with on-site tours, classes, and printed information. The Audubon Society has excellent information to share on birds and will often supply bird feed to nonprofit agencies. Criteria for selecting nature studies or environmental activity areas are the therapeutic value of the activity, the accessibility of tour sites, and whether the activity is age and ability appropriate for the population served.

Summary

In summary, a well-developed horticultural therapy program focuses on the individuals being served, their therapeutic goals, and their individual likes and dislikes. Further criteria in activity selection include

- Therapeutic value
- Type of setting
- Staffing
- Risk management
- Providing continuity
- Planning for the seasonal calendar
- Adaptability
- Recorded success

Following the guidelines outlined in this chapter will help to ensure a safe and successful horticultural therapy program, one in which the participants, gardens, staff, family members, and facility all thrive.

Further Reading

Bell, Virginia and David Troxel. 1999. The other face of Alzheimer's disease. *American Journal of Alzheimer's Disease* 14(1): 60–64.

Carter, Marcia Jean. 2011. *Therapeutic Recreation: A Practical Approach*. 4th edn. Long Grove, IL: Waveland Press.

Catlin, Pam. 2012. *The Growing Difference: Natural Success Through Horticultural-Based Programming*. Create Space, doi: 9781477429662.

Cole, Marilyn B. 2005. *Group Dynamics in Occupational Therapy: The Theoretical Basis and Practice Application of Group Intervention*. 3rd edn. Thorofare, NJ: SLACK.

Geboy, Lyn and Beth Meyer-Arnold. 2011. *Person-Centered Care in Practice: Tools for Transformation*. Verona, WI: Attainment Co, doi: 9781477429662.

Leading Age. http://leadingage.org (accessed July 14, 2015).

Pioneer Network. http://pioneernetwork.net (accessed July 14, 2015).

Wood, Debra. 2013. Building a person-centered culture for dementia care. *Leading Age* 3(5), September/October 2013.

References

Borg, Barbara and Mary Ann Bruce. 1991. *Group System: The Therapeutic Activity Group in Occupational Therapy*. Thorofare, NJ: SLACK.

Buettner, Linda and Shelly Martin. 1995. *Therapeutic Recreation in the Nursing Home*. State College, PA: Venture Publishing.

Rice, Jay S. 2012. The neurobiology of people-plant relationships: An evolutionary brain inquiry. *Acta Horticuturae* 954: 24–28.

chapter four

Working with program participants

Techniques for therapists, trainers, and program facilitators

Karen L. Kennedy and Rebecca L. Haller

Contents

Introduction

This chapter explores techniques that are used to work with participants in horticultural therapy programs. The concepts and methods described are used in many types of health-care and human service fields, and they provide a basic foundation to promote positive outcomes. These approaches are also useful for group leaders, teachers, and program facilitators. By employing them, the therapist/leader will be able to maximize the growth and functioning of those served. Techniques addressed include

- Facilitation and group leadership
- Therapeutic use of self
- Motivation and behavior management
- Training methods
- Adaptation and modification

Facilitation and group leadership

Although not exclusively, most often horticultural therapy is practiced in groups varying in size from 3 or 4 to 15 or more individuals. Effective planning, facilitation, and group leadership techniques are therefore key skills for all horticultural therapists, as well as for leaders of community horticulture programs. Some considerations for planning and leading groups are the populations served, the personalities involved, the purpose and type of group, the activities or tasks to be performed, and the setting of group sessions.

A collection of individuals can be considered a group when they identify themselves as such, interact, and have a shared purpose (Austin 1991). Not all horticultural therapy groups are able to achieve this sense of "being a group" due to factors that interfere with this identification. For example, group members may change frequently, members may only be capable of or allowed minimal interaction, or the group may only meet for one or a few sessions. By consciously cultivating group formation, leaders use a tool that can foster individual growth. Group practice offers a distinct advantage to participants to have their needs met through interaction and relationships with others (Toseland and Rivas 2001, Schwartzberg et al. 2008). In horticultural therapy groups, participants develop relationships with the therapist, the plants, and other group members. This group interaction provides opportunities for support, feedback, social role practice, and a wide array of social skills such as cooperation, communication, and trust (Figure 4.1).

Figure 4.1 Gardening in a group leads to numerous social and emotional benefits for participants. Photo by Rebecca Haller.

Types of groups

This section looks first at the types of groups and how they might be used in horticultural therapy programs. Groups can be classified as *activity* or *support groups*, or a combination of the two (Finlay 1993). *Activity groups* may focus on tasks or social experiences. *Task activity groups* focus on the horticultural activity or task to be performed. Goals may include development of attention span, concentration, completion, cooperation, creativity, or vocational skills. The end product of the group effort is valued and motivating. In *social activity groups*, the emphasis is on leisure interests and social interaction. Participants are encouraged to interact, cooperate, and have fun. *Activity groups* are perhaps the most prevalent type in horticultural therapy practice. *Support groups*, on the other hand, focus on communication and/or psychotherapy. *Communication support groups* aim to share experiences and feelings in order to provide support for group members. *Psychotherapy support groups* focus on reactions, interpersonal skills, and feelings. They may help members address psychological issues and work on personal goals within the group. Examples of these four group types, as used in horticultural therapy, are summarized in the following list:

- Task activity group: group with gardening or vocational production focus
- Social activity group: garden club at assisted-living facility
- Communication support group: group that uses gardening metaphors and shares experiences
- Psychotherapy support group: group gardening used to explore individual feelings and relationships

Figure 4.2 Horticultural therapy groups are strategically designed to meet the needs and goals of group members and overall program.

In practice, horticultural therapy groups may include one or more of these foci. For example, in a "garden club" at a long-term care facility, residents may learn gardening skills, socialize and have fun, talk to each other in supportive ways, and perhaps share feelings as part of the session. In this setting and others, the emphasis of the group may shift, depending on the current needs of the people involved. In effect, the group then functions as a task and social activity group as well as a communication support group (Figure 4.2).

Leadership styles

In order for horticultural therapy groups to be successful, the horticultural therapist must lead the group effectively. The leader's planning, preparation, and style all impact the achievements of the group. The objectives are to create and guide the group in ways that maximize the growth and development of the participants. Leadership styles necessarily vary according to the purpose or type of group, the level of group functioning, and the horticultural activity performed. Various situations call for a diverse repertoire of roles to be adopted. The leader/therapist may act as a teacher, planner, coach, facilitator, or motivator at any given time. For instance, during a support group for people coping with cancer, the therapist assumes the role of a teacher when providing cultural information about tropical plants and techniques for designing and planting a dish garden. As the clients discuss how their choices are analogous to life choices they have made and balance in their own lives, the therapist's role shifts to coach or facilitator of the discussion. Thoughtful consideration and flexibility in adapting to the present need are required.

Three main styles of leadership are commonly described—*authoritarian* (also referred to as *autocratic*), *democratic*, and *laissez-faire* (Austin 1991, Finlay 1993). They represent a continuum of directive to nondirective approaches, each with distinctly different results. The *authoritarian* style is characterized by strong control over the group's actions, with each step directed and closely supervised. At the other end of the continuum, using the *laissez-faire* style involves very little direction by the leader, with group members deciding and initiating action on their own. A *democratic* leader outlines the task at hand, provides choices and some guidance, and encourages the group to discuss how and what to do. While therapists may be most at ease using one of these leadership styles, it is important to become comfortable with the full range of styles in order to facilitate growth and group functioning in the many situations encountered in horticultural therapy programs. Each style offers advantages and may be the appropriate choice for the situations described in Table 4.1.

In choosing a leadership style for each situation, the horticultural therapist should consider the functional level of individuals and the group, the setting, the type of horticultural task to be performed, and the relative importance of the end product. For example, an authoritarian style may be appropriate in a vocational setting with strict quality standards or time constraints, or in a setting where safety issues are paramount. At the other end of the spectrum, a small horticultural therapy support group for people with cancer may benefit from a *laissez-faire* style of leadership with an emphasis on creativity and individual choice.

Fostering functional group behaviors

Group members perform two major functions as described by Austin (1991)—*task functions* and *social-emotive functions*. When used as a description of group dynamics, the term *task function* refers to those activities that help members to achieve their goals, whether these are cognitive, physical, emotional, social, or vocational. The *social-emotive function* of a group is to promote and achieve a positive group atmosphere. In order for groups to perform both of these functions well, the horticultural therapist must provide effective settings, structure, and leadership, and also manage any nonfunctional behaviors that occur. The leader encourages behaviors that help the group reach goals and satisfy group needs (Schwartzberg et al. 2008). Horticultural therapists face many kinds of challenges in leading groups, including those associated with participation levels and disruptive behaviors. Exhibit 4.1 provides some questions to ask when participation is limited or interfering.

Table 4.1 Horticultural therapy practice: Leadership styles

Leadership style	Description	Useful when
Authoritarian	• Directive approach	• The group is newly formed
	• May foster dependence	• There is limited time to accomplish the task
	• Responsibility is with the leader	• High standards exist for task performance
		• The group is very large
		• Participants have limited social skills
		• Participants have cognitive deficits
		• Negative/disruptive behaviors occur
		• Safety is paramount
		• Structure is essential for group function
		• Choices are threatening or overwhelming to clients
Democratic	• Members involved in decision making	• Time allows for discussion to take place
	• Needs time for discussion	• Social skills are key goals of the group
	• Feelings of teamwork	• The group is functioning well
		• It is important to seek full participation
Laissez-faire	• Nondirective approach	• Creativity is emphasized
	• Open and permissive	• Trust and responsibility are goals
	• Client centered	• Group members are able to give and accept social influences
	• Group members initiate action	• The group size is small
	• Emphasizes independence	• Group members need to set their own agenda
		• Standards for the end product are flexible

> **Exhibit 4.1   Horticultural Therapy Practice:
> Participation Levels**
>
> Participation levels can range from nonattendance or lack of partici-
> pation to domination of discussion and attention. Questions to ask
> in these situations include
>
> - Do the facility staff and coworkers support attendance through
>   transportation, scheduling, reminders, and so on?
> - What factors motivate attendance and participation? For
>   example, do the time of day, location, and use of end products
>   contribute to full participation?
> - Are horticultural therapy sessions voluntary or mandatory?
>   Are the sessions tied to a reward system for clients?
> - Do medications affect alertness?
> - Are there other appropriate outlets for personal discussion,
>   counseling, or problem-solving?
> - Are quiet members encouraged to be involved?
> - Do domineering members receive sensitive feedback regard-
>   ing their behaviors?
> - Has the client described a barrier that might prevent
>   participation?
> - Do financial concerns prevent attendance?
> - Are activities appropriate and interesting?
> - Are group members compatible?

Disruptive behaviors may include interruptions, displays of anger
or aggression, or bizarre or socially inappropriate actions. For example,
interruptions could consist of continually asking for help, shouting, walking away, or interrupting another's statements or conversations. Displays
of anger or aggression could include shouts, verbal or physical threats,
destruction of property, or crying. Socially inappropriate behaviors may
involve approaching strangers with hugs, psychotic or delusional epi-
sodes, or overreactions to common situations.

In these and other disruptive situations, the role of the group leader
or horticultural therapist is first to manage the behavior sufficiently
to reduce the real or perceived threat to the perpetrator and the other
group members. The group should be removed if there is immediate
physical threat. Generally, redirecting the individual to appropriate
action is effective. A person with dementia who is experiencing agita-
tion may respond to the therapist's calm voice and a simple gardening
task in which to be engaged. By recognizing signals that indicate that a

group member's anger is escalating, the leader can take action before a crisis occurs. A physically active gardening task such as digging could be assigned. Conversation, appropriate expression of feelings, relaxation exercises, or other techniques could be used. Harvesting leaves or flowers from a scented plant such as lavender, whose fragrance is known to be calming, or a repetitive task such as weeding or cultivating the soil are examples of garden-specific tasks for eliciting behavior change. By understanding each participant, the therapist may also select a task the individual particularly enjoys, such as watering. The garden or greenhouse environment offers many options for calming and redirecting disruptive behaviors.

The therapeutic use of self

The primary components of a horticultural therapy program are the plants and growing environment, client, and therapist. The development of client goals and the plant-related activity selected to facilitate achieving these goals are the foundation of the program. The role of the therapist in this process cannot be underestimated (Schwebel 1993). In fact, the relationship established between the therapist and the client is a crucial element in the overall process.

Therapist's role

The therapeutic relationship differs from a social relationship in that meeting the needs of the client is the basis of the relationship. To be effective, the therapist must be aware of his or her own attitudes, feelings, biases, values, and beliefs, and how these affect interactions with each client. The purpose of the relationship is to address the client's goals. To facilitate this process, the therapist must

- Use effective communication techniques
- Encourage a satisfying relationship
- Provide physical and emotional support
- Facilitate the client's understanding about himself or herself
- Understand the therapeutic process

Other elements of a therapeutic relationship are

- Rapport
- Respect
- Empathy
- Genuineness
- Authenticity

- Trust
- Patience

The use of self

In creating a nonjudgmental atmosphere in which optimal client growth can occur, rapport is built on *warmth* and *acceptance*. Verbal and nonverbal communication reveals the therapist's feelings and attitude toward clients. This includes facial expressions, posture, and tone of voice. Clients connect with therapists who present themselves authentically and who are honest, open, and real with them. Regardless of personal feelings about a client's choices, behavior, personality, or appearance, treating him or her with respect means that the therapist believes unconditionally in the positive value of that individual (Figure 4.3).

Self-disclosure is sometimes useful and appropriate in establishing a therapeutic relationship. When exploring feelings or experiences in common, it may be helpful in expressing empathy. However, sharing personal experiences should be done carefully, not as a venting opportunity, so as not to lose focus on the client's needs.

Figure 4.3 A horticultural therapist at the Margaret T. Morris Center connects with an elder with respect and authenticity.

Creating boundaries in therapeutic relationships establishes trust and safety. Be mindful of the use of touch. Not all individuals interpret touch in the same way and some settings prohibit staff/client physical contact. Be conscious of body position and personal space. Although some work-sites are extremely casual in the horticultural therapy field, a therapist's attire should be benign and professional so as not to distract or interfere with a professional relationship.

Since change occurs slowly and in small steps, exhibiting *patience* is a great form of encouragement and support. This does not mean that one should abandon great expectations, because tolerance of the journey taken to achieve the client's goals can strengthen the relationship and communicate support for the individual's efforts in the process itself.

Finally, two important aspects in the therapeutic use of self to consider are how a relationship with plants fosters growth and understanding in the therapist, as well as how the therapist's relationship with clients impacts his or her sense of self. See Exhibits 4.2 and 4.3 for more information on maintaining a healthy perspective.

Exhibit 4.2   Horticultural Therapy Practice: Techniques in the Therapeutic Use of Self

Ground

It is important to ground oneself when working with others. Without a sense of "grounding," the therapist might need his/her client to achieve a specific goal or act in a certain way in order to feel solid within himself/herself. The horticultural therapist can visualize having a root that extends into the earth. This root can bring balance and ground emotions or concerns that come up for the therapist while working.

Focus

It is important for the horticultural therapist to see the client's work within the context of his or her whole life. It is also important for the therapist to see where the client embodies wholeness despite disability, disease, or disadvantages in life. If the therapist softens his/her focus, as if using peripheral vision when viewing the client, he/she will be better able to hold onto the big picture of this person's life. In doing this, the therapist will be better able to assist the client in accessing this experience as well.

Self-Care

The horticultural therapist's ability to help others is connected to the client's ability to take care of him- or herself. It is important for the therapist to have practices and routines that support replenishment and letting go of the work.

Boundaries

It is important to recognize that the horticultural therapist does not heal the client. The therapist facilitates a relationship between the client and plants that supports the capacity for healing.

Blending

The horticultural therapist's task is to help clients identify goals that fit their life circumstances. This therapeutic contract requires blending with where they are now. The therapist does not predetermine what is best for their clients. The only time that controlling the direction of the work is called for is if a client is in danger of hurting himself or herself or another person.

Closure

Often, the horticultural therapist works with a client for only a short period of time and does not see the results of this work. It is important for therapists to create a place for closure with those they work with. The horticultural therapist can visualize completing a contract with a person. This supports the recognition that his or her life will continue to unfold.

Contributed by J. S. Rice

Exhibit 4.3   Horticultural Therapy Practice: The Gardening Self

(Note: This section is best read in the garden, in the company of plants.)

In a helping profession, the therapeutic use of *self* entails the capacity for self-reflection to know how you are affected in your engagements with other people. In horticultural therapy, the therapeutic use of self additionally draws on your experience of yourself as a part of the natural world. Take a moment after reading this, set down your book, and feel your connection to this (mother) planet earth.

Imagine or intend to extend your energetic roots down into the soil that you are standing or sitting on. What do you experience in this moment?

Extend your awareness to the plants that are surrounding you. As you breathe out, observe that you are feeding the plants. As you breathe in, observe that the plants are feeding you. What do you notice when you do this? Isn't it interesting that the Latin and Old French root for *inspiration* refers to our lungs breathing in, as well as our experience of being inflamed or catalyzed by spirit?

Think of gardening. How has it influenced your life outside of the garden? Have you considered that each person you work with has many seeds within? What have you learned that will help them grow?

In the garden, we learn that we are interdependent. The garden does not grow solely by our effort or force of will. Rather, the garden represents our co-creation with the plants, the winds, the rains, the insects, birds—with all of life. How might this recognition shape your work as a horticultural therapist?

Contributed by J. S. Rice

Communication techniques

Attention to the elements of a therapeutic relationship mentioned previously establishes the foundation for effective communication. Helping clients to become more at ease with plants if they fear that they have no "green thumb" is often another helpful first step. Pictures, demonstrations, and the plants themselves facilitate the process for most people—especially if memory, abstract thinking, or unfamiliarity is an issue.

During discussion, ask open-ended questions to elicit a more lengthy response and encourage descriptions of thoughts or feelings. "Why" and "how" questions tend to put people on the defense. When appropriate, draw on examples from the plant world to illustrate points, make abstract concepts more concrete, or provide cues as in Exhibit 4.4.

Humor is often useful to create a positive, relaxed atmosphere, particularly with clients the therapist knows well. An example of the use of humor is shown in Exhibit 4.5. It is a good technique to respectfully incorporate when a task may be frustrating, to alleviate anxiety or nervousness when trying something new, or when teaching stress-reduction coping skills. Humor is not particularly constructive with people who have difficulty thinking abstractly or who are coping with major depression.

**Exhibit 4.4   Horticultural Therapy
Practice: An Oak Tree Cue**

A client with a traumatic brain injury who had difficulty with awareness of his body posture used an oak tree as a cue. This client over
time would slowly begin leaning so far over he would eventually
fall out of his chair. An acceptable, age-appropriate cue to correct his
posture was to simply say "Like an oak tree." He selected an oak tree
because to him it symbolized strength and uprightness. It was also
a more dignified cue than hearing the therapist say "Sit up straight"
frequently.

**Exhibit 4.5   Horticultrual Therapy Practice:
Plants and Humor**

Humor can be interjected into horticultural therapy sessions through
words or the planting project itself. Consider purchasing or creating containers with faces and selecting a variety of plants to grow
as hair, or add a nose and glasses to a plant or pot. Make up your
own common names for succulents. Make seed packet cards with
original puns such as beet seeds—"my heart beets for you"; lettuce
seeds—"lettuce be friends." (Figure 4.4)

Figure 4.4 Humor is employed at Skyland Trail through the creative use of
plants and gardening materials. Photo by Christine Capra.

People respond to plants in fantastic ways. They are encouraged to try harder, are distracted from limitations, are motivated by previous (as well as the anticipation of new) experiences, and can achieve a myriad of physical, psychosocial, and vocational goals. It is the relationship with the therapist that connects the optimal horticultural experience to the client's best interest.

In summary, the primary goals of the therapeutic relationship are to

- Facilitate communication
- Facilitate acquiring new skills
- Encourage and support efforts made toward established goals
- Promote independence

Motivation and behavior-management techniques

Horticulture is a powerful therapeutic tool for individuals who have an interest in and enjoy the beauty of plants and flowers. People respond to the sensory stimulation inherently involved in interacting with plants and flowers (Haas and McCartney 1996). For others, the usefulness of plants for food, crafts, decorating, and so on has appeal. Horticultural therapists use the people–plant response to motivate clients to participate in both group and individual sessions. Horticultural therapy sessions are an ideal venue for positively impacting behavior because of the relaxed and non-defensive physical and mental state that participants tend to exhibit during horticultural therapy activities. In rehabilitation settings, for example, physical and occupational therapists and rehabilitation nurses, as well as professionals working on spiritual, behavioral, and chemical dependency issues, find their work to be positively impacted by the participant's response to the horticultural therapy program (Murrey et al. 2001). When clients are less agitated, defensive, or tense, they are more receptive to challenging therapeutic interventions.

Positive reinforcement

By their very nature, plants respond to care. The direct causes and effects of watering, fertilizing, pruning, and grooming lead to healthier plants. New blossoms and fruits are tangible rewards that participants can identify and feel good about (Figure 4.5). Therapists use this concept to reinforce desired behaviors such as increased

- Frequency and level of participation
- Standing tolerance
- Tolerance of tasks
- Positive interactions with others in the group
- Refocus on health and wellness

Figure 4.5 Growing food for sale offers work that is both genuine and rewarding for program participants. Photo by Rebecca Haller.

In groups, the therapist can manage behavior and encourage pro-social behavior by giving group participants tasks they enjoy as positive reinforcement. For example, a participant in a prevocational setting has trouble working without arguing with others. If he follows through with weeding for 15 minutes without arguing with another participant, he will be given the task of watering at the end of the session, which he particularly enjoys. An important part of the process is bringing the participant's attention to the desired behavior and the reward.

The growth that participants see within the garden or growing space is used to reinforce progress on goals and desired outcomes. This awareness of growth is an effective strategy to build confidence and self-esteem. Participants may even exceed their goals and surprise themselves at what they can achieve. The process continues to build as participants accept new responsibilities and advance their progress on cognitive, behavioral, and mental health goals. It is useful for horticultural therapists to verbally recognize growth and change of the participant or to simply reinforce progress with verbal praise such as "good job."

Horticulture as a motivator

Growing plants is an experience that participants are likely to be familiar with before they are part of a therapy program. It is viewed as a meaningful activity, especially for those who have lost their independence as a result of an injury, illness, or living situation. Taking plants back to their living space provides participants with a sense of control over their environment. Participants may find their own meaning and motivation in the planting process or activity by drawing comparisons or analogies to their own life situations. Pruning the plant to encourage new growth and free it from tangled overgrowth illustrates the pruning that one can do in one's own life to be free from any parts that are overwhelming.

In some programs, being allowed to participate in the horticultural therapy program is a reward for good behavior. In many correctional facilities, for example, inmates must demonstrate nonviolent behavior and adherence to the rules before being admitted into the program. This behavior must continue as they learn new vocational and coping skills that will enable them to be successful citizens when released. Programs that facilitate behavioral change also help clients identify obstructions to a healthier and better life. With the support of the horticultural therapy group, clients learn how to cope with these barriers and begin to weed them out.

The basic activity of plant care may be used to provide reality-based situations to motivate self-improvement. For example, if poor behavior means that a client is confined to his or her living quarters and is not allowed to attend horticultural therapy sessions, rather than "rescuing" the plants, the therapist may allow the plants to be neglected. This provides an opportunity to learn cause and effect through natural consequences.

In programs where participants grow and sell plants, the customers provide the positive reinforcement. Participants are often able to share valuable information about plant care with their customers, who may be peers, staff, or community members. This knowledge and the ability to produce something that others esteem and are willing to pay for are empowering and motivating. Participants shift from their role of care-receiver to caregiver by becoming the ones providing both information and a valued product to others.

Tips and techniques

Selecting the most appropriate plants, tasks, and activities is key to maximizing the desired behavioral responses. It is important to match the plants and activities to the participants' preferences and interests. A short interview or survey can yield favorite flowers, colors, or fragrances as well as special interests such as cooking or specialty plants and gardening history. For some participants, familiar plants, flowers, and fragrances are comforting, reinforce self-confidence in new and unfamiliar situations, and trigger reminiscence. New experiences and intriguing plants captivate others. Understanding a client's attitude toward plants and gardening as well as how he or she feels about a living environment or treatment program is necessary in choosing the best approach for the individual.

Be mindful of gender and ethnic backgrounds. For example, after a presentation on herbs in a mixed gender group, it may be more effective to

Figure 4.6 The greenhouse setting provides a safe and motivating environment in which to try new skills. Photo by Rebecca Haller.

plant a herb container garden rather than make fragrant bouquets (such as tussie-mussies or nosegays). Women tend to be more comfortable making bouquets of flowers. The activity of planting a herb container garden is more gender neutral and supports the planned discussion.

Special considerations may exist that are unique to facilities or the conditions of the people served by the program. For example, some people may be sensitive to fragrance due to pulmonary issues or chemotherapy, or some may tolerate natural plant fragrance better than synthetic smells. Others may lack the sense of smell altogether due to a brain injury.

Fortunately, the diversity of the horticultural world offers a plethora of plants and associated tasks that appeal to a wide range of interests and abilities. As described under "Adaptation and Modification," activities and tasks can be graded to increase in difficulty or complexity as goals are achieved. The safe environment established by the greenhouse or garden may be ideal for learning new techniques. For instance, participants may be more motivated to learn compensatory strategies for memory or cognitive deficits in the safety of the greenhouse or garden setting. Once comfortable with the strategy, it can be transferred to other vocational or living situations (Figure 4.6).

Training

In all types of horticultural therapy programs, the horticultural therapist is called upon to provide teaching or training to individuals and groups. The training may involve specific horticulture skills of various complexities, from planting seeds to pruning fruit trees, or may be a simple instruction such as "find a node on the coleus plant." Or the aim may be to help an individual learn a social skill, a self-help behavior, or other activity. This section focuses on those teaching/training skills that are used to lead individuals and groups in the performance of horticulture tasks or activities. The therapist's role as trainer or teacher requires an in-depth understanding of the task or activity to be performed and the type of learning desired, as well as specific techniques to support or elicit competent performance.

Task analysis

Understanding the task or activity involves a process called *task analysis*. (The terms *task* and *activity* are used interchangeably in this chapter to refer to the horticulture action to be performed by the client.)

The process involves looking at

- The steps to successfully perform the activity (Lamport et al. 2001)
- The materials, equipment, and facilities required
- Possible adaptation or modification to materials, instruction, or environment

It may also include

- Analyzing the skills necessary to perform the task
- Looking at the cultural, age, and developmental appropriateness of the task
- Identifying the therapeutic qualities of the task

This section focuses on the necessary steps to do the task, including developing an awareness of what is entailed and the concrete actions required.

The analysis begins with the performance of the task. More experienced gardeners often overlook this essential step. While familiarity with the task or activity at hand increases confidence and contributes to horticultural success, leaders also need to be aware of the meaning, effects, and demands of each activity from the client's perspective. Furthermore, it is important to identify a preferred method for accomplishing the task, as there may be several acceptable approaches to any particular gardening task.

First, as the task is performed, the therapist should notice and record the following:

- Any thoughts or feelings evoked while doing the task
- The physical and cognitive requirements to complete it
- Any emotional response upon completion

This helps the therapist/leader to understand some inherent qualities of an activity and possible meanings it may have to program participants. Understand that background, culture, and health status will all affect an individual client's reaction to and experience of each activity.

Second, record the materials, equipment, and space used in the activity.

Lastly, identify the actions necessary to perform the activity. Write out the sequential steps that were completed in concrete, observable terminology. In other words, describe what was done that could be directly observed by another. Include as much detail as necessary to clearly show what took place. It is helpful to watch someone else perform the same task and use the observations to modify the list of sequential steps (see Exhibit 4.6 for an example of a basic task analysis that lists the materials and steps necessary for performing a horticultural task).

Using this example as a starting point, the therapist alters the task analysis to increase its usefulness in an actual horticultural therapy setting, adding steps and materials that are specific to the person, place, and situation. The basic performance steps may be modified to allow a client to successfully complete the activity. Changes in the setting, tool

---

Exhibit 4.6   Horticultural Therapy Practice: Task Analysis

Task: propagation of indoor foliage plants by stem tip cuttings.
Materials: stock plant, pruners, container(s) filled with potting soil, and water.
Steps:

1. Find direction of growth on stock plant and locate the tip.
2. Measure an index finger length from the tip of the stem.
3. With the pruners, cut the stem at the point located about one finger length from the tip. Be sure that the cutting point is located at least below the third node.
4. Remove leaves from the lower one or two nodes.
5. Push a finger into the center of the potting soil in a container to make a hole about two inches deep.
6. Place the cutting in the hole, with the bottom nodes (those that are leafless) below soil line.
7. Gently firm the soil around the cutting to hold it upright and ensure soil contact with the nodes.
8. Repeat for the desired number of cuttings.
9. Water.

and materials selection, amount of detail or number of steps to perform, degree of self-initiation required, or time allowed may be necessary for success. Based on knowledge of the client's functional level, the therapist should aim to encourage the most independent performance possible and should gradually reduce the modifications or supports as the client progresses. Modifying the basic task analysis results in an individualized approach for training and performance. Further information on adaptation and modification is provided later in this chapter.

A clear outline of performance steps, modifications, and supports makes it possible to have consistent expectations of the client. It even allows multiple trainers to maintain this consistency. Additionally, the task analysis may be used to document performance on each step, create a record of progress, and identify the need for further training or adaptation (see Exhibit 4.7 for another example of how the sequential list of steps might be used).

Exhibit 4.7   Horticultural Therapy Practice: The Importance of Following Directions

Individuals with traumatic or acquired brain injury in a vocational training program may have a goal to follow multistep directions. In this case, it is important for the therapist to determine the best method for accomplishing the task, give clear and succinct instructions, and provide any compensatory strategies necessary for accomplishing the task. A checklist of the steps is an example of a compensatory strategy for short-term memory loss.

For example, Adam may decide that rather than filling a pot, selecting a rooted cutting, transferring it to the pot, and adding the remaining soil as instructed, he would prefer to do the task assembly-line style (filling many pots at a time, etc.) While his method would still achieve the desired end result, he would not meet his goal of following directions. In this case, it would be up to the therapist who is filling the role of work supervisor to decide how to manage this individual. Requiring the client to check with the supervisor before making procedural changes reminds the client that following directions is important to succeed in getting and keeping a job. It also addresses communication skills and appropriate work behaviors.

Note: For other types of groups, following directions exactly as given may not be as important as comprehension and the client's satisfaction with the final product.

Types of learning

In addition to analyzing the task or activity to be performed, it is helpful to be clear about the type of learning required of the client. Is the aim to learn facts, concepts, principles, procedures, interpersonal skills, or new attitudes? Table 4.2 shows teaching strategies for each of these types of content (Kemp et al. 1998).

To determine the types of learning and the techniques to use, the therapist also considers the therapeutic goals as well as the abilities of participants.

Goals and objectives

Types of learning and training techniques used should correspond with the objectives of the individuals served. For example, in a vocational horticultural therapy program, the client may be called upon to engage in any of the learning types shown in Table 4.2. More specifically, he or she may learn the procedure to transplant seedlings or to work cooperatively on a task—two very different types of learning. The focus depends on the person's current functioning and treatment objectives.

Skills, abilities, and experiences of individuals

The type of learning to be expected is also based on the level of experience of those involved. For example, facts generally need to be understood prior to learning the principles associated with them. Similarly, it is useful for a client to know facts (such as identification of plant parts) before teaching him or her a procedure (e.g., taking stem cuttings of houseplants).

Table 4.2 Horticultural therapy practice: Teaching strategies for various content or performance

Content	Teaching strategy	Horticultural example
Fact	Show, practice, and rehearse	Ripe "Early Girl" tomatoes are red.
Concept	Show, describe, and organize	Identify tomatoes ready for harvest.
Principles	State principle or rule	Plants need healthy roots to grow.
Procedure	Demonstrate, organize, elaborate, and practice	Transplant seedlings to pots.
Interpersonal	Model, imagine, and rehearse	Cooperatively mix soil and fill flats.
Attitude	Model, imagine, and rehearse	Gardening helps me to reduce stress.

Strategies for teaching procedures

The basis of horticultural therapy is active participation in a horticulture activity or task. The client performs basic or complex procedures through involvement in the therapy session. Therefore, horticultural therapists frequently are called upon to teach procedures. As seen in Exhibit 4.8, to teach procedures, the therapist most often begins with a demonstration or modeling of some sort. A description of this common training technique, as well as verbal and physical methods, is shown in Exhibit 4.8 in order to give the therapist options for meeting the training needs of various participants (Callahan and Garner 1997).

Depending on the setting and population served, horticultural therapists may find each of these techniques useful. The techniques may also be combined, as when verbal instructions are given along with a

---

**Exhibit 4.8   Horticultural Therapy Practice:
Training Techniques**

Demonstration: shows the object or performance

- Uses a common and natural way of teaching or training
- Shows acceptable methods for doing a procedure
- Requires that the client observes carefully
- Helps learning and self-confidence to see the end result of the product (activity)
- Is often accompanied by verbal information to explain details or context
- Initiates the activity—usually occurs before the client begins

Verbal: uses spoken instructions, directions, or reminders

- Allows the therapist to explain details or the bigger picture
- May cue or prompt the client to start the next step or to correct performance
- May create an overload of information, particularly in noisy settings or when mixed with conversation

Physical cue: uses a physical motion to teach or prompt

- Includes
  - Gesturing: pointing to object or area to prompt next step
  - Modeling: miming the desired motion without words—simulating the necessary movements

- Effective in noisy environments or when language barriers exist

Physical assist: using hand-over-hand motions to assist the client in performing the task

- May be useful with people with severe cognitive, sensory, or physical disabilities
- May be useful when safety is an issue
- Only use if necessary, as this method allows for less independence of the client
- Be gentle and respectful; do not force; gauge the reactions of the client (Figures 4.7 and 4.8)

Figure 4.7 A horticultural therapy student tries out a physical cue during a class exercise. Photo by the Horticultural Therapy Institute.

Figure 4.8 Hand-over-hand assistance is desirable when other techniques are insufficient and the client is receptive to its use. Photo by Christine Capra.

demonstration or physical cue. Generally, the therapist should choose the technique or assistance that is least restrictive or interfering in order to be respectful and to encourage independence. Other factors to consider when choosing a training technique include individual learning styles of clients, safety issues, and the relative importance of quality plant production or garden appearance.

Learning styles

Leaders may accommodate various learning styles by using an assortment of teaching techniques. Individuals tend to prefer or focus on one means of gathering and processing information. They may prefer visual, auditory, or kinesthetic processes to learn and retain information. Similarly, people process information in various ways—learning best by considering broad concepts or in sequential fashion, by intuition, or by observation, and so on.

Safety issues

Choose training techniques that maintain a safe working environment. For example, physical assistance is commonly necessary to teach a four-year-old visually impaired child to cut houseplants for propagation (Figure 4.9).

Figure 4.9 Use of sharp tools may necessitate a more restrictive or supportive training technique to ensure safety. Photo by Christine Capra.

Importance of horticultural success

Choose training techniques based on production demands and needs for attractive well-kept garden environments. In other words, weigh the relative importance of the activity process and the end product against the type of horticultural therapy program offered. Keep in mind that a well-kept garden and successful plant growth are generally desirable and motivating to participants.

Adaptation and modification

To meet the varied needs, goals, and abilities of the people served in horticultural therapy programs, leaders adapt methods and modify the activities to be performed. As previously stated, the activities planned must be accessible to the participants and appropriate for their ages and abilities. The use of horticulture allows for many gradations and methods to accomplish any given task. Adjustments to an activity are planned for in advance, based on knowledge of each client, and may be indicated in a task analysis. However, in practice, the therapist is typically also required to modify aspects of the task on the spot in response to the individual functioning seen during the activity. For instance, a person may need more instruction than usual due to medication or illness. Someone who has just experienced an emotional event might be overwhelmed by distractions or noise in the greenhouse activity space. Also, since most horticultural therapy sessions take place in groups, variations in mood, cognitive functioning, physical abilities, and social skills are regularly encountered. Flexibility and attentiveness to the people served are essential talents of horticultural therapists.

Adaptation

Adaptation involves changing the means to accomplish the task in order to enable the client to be successful. It includes adaptations to tools, the environment, the client's position, and the instruction provided (Hagedorn 2000). The outcome or end result of the activity is not changed (Lamport et al. 2001) (Figure 4.10).

For example, in the task analysis shown in Exhibit 4.6, the end result of the task itself is that a cutting is correctly planted and watered. The use of round-tipped scissors as a *tool* adaptation may enable a young child to be successful and safe. For someone with autism, changing the *environment* so that the entire sequence of steps is performed individually in a quiet section of the greenhouse or activity room may be necessary in order to enable the individual to focus on the task at hand. (Although beyond the scope of this book, garden design is a type of modification that is also often necessary for horticultural therapy programming.) Performing the task in a standing position, such as at a potting bench, is an adaptation of *position* that provides an opportunity to work on an objective to increase standing endurance. The type and frequency of *instruction* given can also be adapted to the needs, abilities, and goals of program participants. Single-step verbal directions may be necessary for a client who is unable to remember or follow a sequence of steps. Exhibit 4.8 describes other types of instructions used for training that may be considered. In these examples, the purpose of adaptation is to alter how the task is to be accomplished by the individual client—matching methods with abilities.

Figure 4.10 Sometimes an adaptation is as simple as adding a tool to enable independence and horticultural success. Photo by Anna Terceira.

Modification

In addition to altering the methods used to accomplish a task, the task itself can be modified or graded to increase or decrease the performance demands placed on the client (Hagedorn 2000). Incremental modifications may be made as the client progresses or declines, facilitating full participation and goal achievement. Possible modifications to or grading of the task of taking stem tip cuttings (Exhibit 4.6) are

- Changing the *length of time* allowed for the activity or the number of cuttings expected in a given time period
- Modifying the *difficulty* or complexity by preparing the cuttings for potting before the session
- Adjusting the *demands* of the activity (could be physical, sensory, social, perceptual, or cognitive) by altering the social aspects of the task—requiring group members to share tools and materials

Again, the idea is to modify the task itself to fit the needs and abilities of the person performing the task.

Recommendations

Additional principles to consider when choosing adaptations and modifications to horticultural activities are *normalization* and *empowerment*. The activity should be offered in a manner that is normal and real to the participants. For example, conduct sessions in the garden itself whenever possible (Figure 4.11). Be sure that activities are age appropriate and

Figure 4.11 A horticultural therapy session in the garden offers a myriad of opportunities and challenges that are empowering to program participants. Photo by Anna Terceira.

Exhibit 4.9   Horticultural Therapy Practice: Promoting Independence

Promoting the highest degree of independence possible is a goal of most horticultural therapy programs. Techniques vary depending on client group and type of program, but the results empower, motivate, and improve self-confidence and self-esteem. Some examples are

- Provide choice within an appropriate scope that can provide some control over clients' living environments and improve their satisfaction with their living arrangements. This might mean providing a choice of three ribbon colors or letting them decide which tomato varieties to plant.
- Supply the materials necessary for the project; however, rather than having all of the materials in front of each person, encourage participants to scan the room for items they need, initiate a request for an item, or get it themselves.
- Encourage problem-solving skills by allowing the group to make choices about how they accomplish a task, items to add to the garden, or ways to address situations (e.g., aphids on basil plants).
- Have garden containers or beds at different heights to provide gardening opportunities for those who need to stand, sit in a chair or wheelchair, or sit on the ground.

fit within a normal routine for those served. Clients can be empowered to perform independently by offering choices, garden features that are enabling, appropriate challenges, and opportunities to participate in planning efforts (see Exhibit 4.9 for examples of how to promote independence and empowerment).

Therapists should take care to modify activities only as necessary, allowing clients to experience the essence of the gardening experience with minimal intrusion (Buettner and Martin 1995). Modifications should also help individuals and groups to meet stated goals and objectives. Exhibit 4.10 and Table 4.3 show how a basic activity might be changed for specific populations and purposes. Examples of techniques for using horticultural therapy to address various issues faced by program participants are presented in Appendix IV. These approaches can be applied in many different settings beyond those listed. For example, under the category of major depression, an issue identified is low self-esteem, a concern that may be present across many populations and settings. Note that each idea is intended to address an issue faced by the client.

Exhibit 4.10   Horticultural Therapy Practice: Adapting and Modifying Activities

To make a tabletop topiary, clients create their own frames by manipulating 12-gauge vinyl-coated copper wire into shapes. The frames, soil, and ivy plants are placed in appropriately sized pots. The project is completed when the ivy plants are gently twisted around the frame. Notice how the basic activity is modified and/or adapted for the following groups.

Elders in a nursing home

The session goals are to stimulate social interaction between the residents and provide opportunities for reminiscing and creative expression. Residents are directed to create shapes for use during the winter holidays such as evergreen trees, wreaths, stars, and so on. In addition to long ivy plants, residents are provided with ribbon and dried flowers for additional decoration. The discussion includes holiday decorating and traditions. For individuals with arthritis or weak grip, use an easier-to-bend wire such as 14- to 16-gauge wire.

Adults with chronic illness

The focus of this support group is to facilitate a discussion that helps participants identify areas in their lives that need support and discuss strategies for shaping their lives in new directions. Clients are directed to create shapes that symbolize the growth and change they intend to work toward. Provide wire cutters and accessories such as dried flowers, pipe cleaners, craft sticks, and ribbon to enable maximum creativity.

Adults with intellectual disabilities

In a vocational program, adults will be creating topiary wreaths to sell during a spring sale. Clients work on following directions and consistency for a saleable product. A template or form will be used to create frame uniformity. A pictorial checklist taped to the work surface will help clients create the frames and plant the topiaries.

Table 4.3 Horticultural therapy practice: Adaptations

Disability/illness of client	Adaptations
Intellectual or developmental disability	• Limit the number of plant choices. • Use sturdy rather than fragile plants.
Hip or knee replacement	• Raise the planting surface to encourage standing. • Provide a variety of plants and accessories to choose from to encourage attention to the process.
Stroke, left side affected	• Place some materials on the participant's left side to encourage visual scanning of the workspace and use of left hand. • Provide several choices to encourage planning and organization.
Chronic illness	• Provide a wide variety of plant choices to enable participants to select plants that represent their current short-term and long-term goals for health management.

Note: Using the activity of planting a dish garden, this example illustrates how the same activity can be adapted based on treatment needs or to enable independence.

Summary

In order to maximize the potential goal attainment of program participants, horticultural therapists need to acquire and practice an array of skills and techniques to employ in day-to-day interactions and programming. This chapter illustrated

- Facilitation techniques
- Therapeutic use of self
- Motivation and behavior management
- Training methods
- Adaptation and modification

All are vital tools used by horticultural therapists for effective treatment and positive outcomes.

References

Austin, David R. 1991. *Therapeutic Recreation: Processes and Techniques.* 2nd edn. Champaign, IL: Sagamore Publishing.

Buettner, Linda and Shelley L. Martin. 1995. *Therapeutic Recreation in the Nursing Home.* State College, PA: Venture Publishing.

Callahan, Michael J. and J. Bradley Garner. 1997. *Keys to the Workplace: Skills and Supports for People with Disabilities*. Baltimore, MD: Paul H. Brookes Publishing.

Finlay, Linda. 1993. *Groupwork in Occupational Therapy*. Cheltenham, UK: Stanley Thornes.

Haas, Karen L. and Robert McCartney. 1996. The therapeutic quality of plants. *Journal of Therapeutic Horticulture* VIII: 61–67.

Hagedorn, Rosemary. 2000. *Tools for Practice in Occupational Therapy: A Structured Approach to Core Skills and Processes*. London: Churchill Livingstone.

Kemp, Jerrold E., Gary R. Morrison, and Steven M. Ross. 1998. *Designing Effective Instruction*. Upper Saddle River, NJ: Prentice-Hall.

Lamport, Nancy C., Margaret S. Coffey, and Gayle I. Hersch. 2001. *Activity Analysis & Application*. Thorofare, NJ: SLACK.

Murrey, Gregory J., Ann Wedel, and Jeff Dirks. 2001. A horticultural therapy program for brain injury patients with neurobehavioral disorders. *Journal of Therapeutic Horticulture* XII: 4–8.

Schwartzberg, Sharon L., Margot C. Howe, and Mary Alicia Barnes. 2008. *Groups: Applying the Functional Group Model*. Philadelphia, PA: F.A. Davis Company.

Schwebel, Andrew J. 1993. Psychological principles applied in horticultural therapy. *Journal of Therapeutic Horticulture* VII: 3–12.

Toseland, Ronald W. and Robert F. Rivas. 2001. *An Introduction to Group Work Practice*. 4th edn. Needham Heights, MA: Allyn and Bacon.

chapter five

Planning horticultural therapy treatment sessions

Karen L. Kennedy

Contents

Introduction

This chapter focuses on integrating the components of a horticultural therapy treatment session into a cohesive and meaningful program element. Attention to strategic planning enables focus both on the big picture as well as on small details for positive outcomes. Using a tool such as the session plan in Appendix II facilitates the process. Some of the elements of a horticultural therapy treatment session have been discussed in previous chapters, and many are integrated into this more detailed discussion about session planning.

The horticultural therapist begins with what is known about the program and its participants, to

- Understand unique characteristics and circumstances
- Review individual treatment goals and objectives
- Identify the type of group (vocational, therapeutic, or wellness)
- Identify the treatment space(s)

Figure 5.1 An outdoor school garden is an effective environment for horticultural therapy programming. Photo by Loredana Farilla.

Certainly, the season and available plant resources are a large factor in activity selection. Identify

- A seasonally appropriate task and/or theme
- Whether the activity is a step in a bigger project or a stand-alone task
- Safety issues with the task and facility specific to the participants
- The number of staff and/or volunteers available to match the assistance level necessary to maintain safety

When available, tasks in the garden and greenhouse provide real opportunities for activity that may be highly motivating to participants. Real versus made up "work" is an especially important element in vocational programs, for example. Plant-rich environments also provide a peaceful and soothing treatment space for all groups (Figure 5.1).

Successful treatment session planning is done through the lens of all of these elements.

Goals and documentation as a basis for sessions

Participant treatment needs are articulated and recorded in the form of goals, objectives, and documentation. An understanding of the desired participant outcomes enables the horticultural therapist to determine the most appropriate way to facilitate the process. A focus on participant goals ensures that the task or activity serves primarily to support participant outcomes rather than becoming the purpose of the session itself.

As discussed in previous chapters, a treatment team often establishes participant goals (see Chapter 2 on establishing participant goals and objectives). Effectively planned sessions must be deliberate about participant objectives. Objectives are essential in defining the specific steps to achieve the participant's goal. In other words, the objectives create a path or roadmap of observable behaviors that demonstrate goal achievement. These observations are reflected in the documentation. Documentation is important since this is the tool that signals the time to reassess. As new challenges are identified or objectives are achieved, new objectives—and perhaps goals—are established.

Horticultural therapy programs in less clinical settings may have very general participant goals. Wellness programs often have participants that are self-referred, with no formal treatment goals. In these cases, the horticultural therapist can engage the participant in establishing personal goals and objectives. With guidance, participants can determine achievable and measurable steps toward the change they would like to see in themselves. Drawing awareness to the therapeutic aspect of the session sets the stage for mindful participation. Participants become more invested in the process, and the program has built-in motivation when they can see personal changes and progress.

Selecting a seasonally appropriate task that enables a range of treatment goals and objectives to be met is, of course, the challenge of planning an effective horticultural therapy session (Figure 5.2). (Chapter 3 covers strategies for choosing tasks and activities in horticultural therapy practice.)

Evidence-based practice

To ensure a high-quality horticultural therapy program, another good tactic in session development is to incorporate evidence-based practice. This involves keeping up to date with the latest techniques that have been documented as effective. Using "tried and true" methods assures the therapist and those being served of a greater likelihood of success. This attention to effective methods adds credibility to the program and to the profession. One primary source for evidence-based information is the American Horticultural Therapy Association (AHTA). AHTA produces the *Journal of Therapeutic Horticulture* and offers online networking, as well as conducting an annual conference. Other resources are regional conferences on horticultural therapy and related areas, colleges and institutes that provide horticultural therapy course work, and colleagues in the field. The following list offers reasons to use evidence-based practice:

- Ensures quality for consumers
- Keeps one up to date to use the best practices available

Figure 5.2 Choosing plants and layout of beds prior to the spring planting season encourages "ownership" and buy-in by participants. Photo by the Horticultural Therapy Institute.

- Uses tested and true methods
- More cost-effective for consumers and third-party payers
- Adds credibility to the horticultural therapy profession
- Self-pride from knowing you are doing the best possible work
- Increases reimbursement, which translates into the availability of more horticultural therapy jobs
- Eliminates methods that are less effective
- Helps shape the future of the horticultural therapy profession

Therapeutic interventions

Session structure

The organization or structure of a session depends on a few primary variables. The type of program (vocational, therapeutic, or wellness), type of task, and participant objectives factor into how a session can be facilitated. This structure can vary from session to session with the same group, or it can remain consistent. Examples of ways that a session can be structured are

- Group introduction and discussion → task → conclusion/wrap-up
- Group introduction → group task, assembly line, or all do each step → wrap-up
- Individual introduction/directions → individual workstations → group or individual wrap-up

Leadership style does change with each of the examples listed and enables the therapist to utilize the structure to best suit the type of group and the participants' functional level, objectives, and task (see Chapter 4 for more discussion on types of groups and the use of leadership style).

The following is an example of a choice between two different session structures for the same horticultural task. Having a group discussion to develop an action plan for a task, and then actually completing the task, benefits participants with goals of problem-solving and communicating appropriately in a group setting. A group focusing on following instructions and staying on task could do the same task without the planning discussion. By the end of the session, both groups could have completed the same task, while working on different skills and objectives.

Session process and plans

Notice that each of these examples includes a specific opening and closing of the session. The opening includes an introduction to the topic, task, and perhaps to the participants and therapist. The use of intentional language during the opening sets the tone for the session. The therapist engages participants with an overview of what to expect to help them feel part of the process, increase comfort, and decrease anxiety. It is often appropriate to include a review of the therapeutic purpose at the beginning of the session, while being observant of participant privacy. Conclude with a summary of group and/or individual efforts and accomplishments (as appropriate). This type of wrap-up sets the stage for the next session from both a horticultural and therapeutic perspective, as well as providing continuity and purpose for participants.

Whether the horticultural task for the session is based in the garden or is an activity such as potting up a dish garden, there are generally multiple ways to accomplish most tasks or projects. The therapist should think through the task steps or process to identify needed modifications for individuals (see Chapter 4). Identify the best way to communicate the task steps and topic(s) of the session. Are there any secondary tools or supplies that would facilitate the process?

Setting up the treatment space can be done when the dynamic of the session has been thought through. Consider the best seating arrangement for effective communication, accessibility, and comfort. Depending on the session focus, participants may need to be able to see each other or work together as a group or with partners, or may need individual workstations. The ideal arrangement also ensures that the therapist can easily observe all of the participants (Figure 5.3).

Another strategy in effective planning includes organizing the session and carefully planning the procedures. Presenting clearly identified procedures in a well-organized session makes it easier for participants to function independently. Using a planning template such as in Appendix II, the horticultural therapist can map out the session plan, including appropriate training techniques, discussion questions, and intentional language. It is a good idea to create a time line to ensure time is managed appropriately throughout the session. By having a detailed plan in place, the therapist is better able to react to necessary changes without losing focus.

Finally, a natural result of creating a plan is a defined set of expectations. Yet, unpredictable variables occur, such as the weather, which can influence plant growth rates as well as participant experience in the garden. Past or recent experiences that participants bring to the session can have an impact on how a participant responds to the session in both positive and negative ways. Horticultural therapy sessions can build on the knowledge that participants have gained during other times in their lives,

Figure 5.3 This strategically organized activity space inherently encourages students to work as partners to accomplish the horticultural task, allowing for social skill development in a natural environment. Photo by Loredana Farilla.

which they may choose to share with the group in a positive manner. Having no past or recent experience with plants and gardens may cause participants some concern, or they may approach the session with gusto since there is no history of failure.

The intrinsic sensory nature of plants is a wonderful tool for use in horticultural therapy sessions (Figure 5.4). Be aware that both positive and negative associations may be attached to plants. For example, during a session in which group members made tussie-mussies (small nosegays made from herbs and flowers), one of the flowers triggered a memory of corsages that the husband of a participant with mid-stage dementia had given her while they were dating. While revisiting happy memories was the intent of the session, and the corsages were a happy memory, remembering that her husband was now deceased was a sad memory to surface (see Exhibit 3.1 for further discussion about response to fragrance). While realistic expectations are important, it is equally important to expect the unexpected!

Figure 5.4 Sights, fragrances, and textures abound when working with plants. Photo by Rebecca Haller.

Material and resource needs

Horticultural therapy programming is often rooted in seasonally appropriate tasks and projects. This mind-set takes advantage of what is happening in the garden on-site, utilizes materials locally available, and often meets the interest of participants. Use the planning template (Horticultural Therapy Session Plan in Appendix II) to list both materials that need to be purchased or acquired from outside the treatment space as well as materials that need to be gathered from within the treatment or storage space. By dividing the materials into these two categories, it is easier to be sure that nothing is overlooked in advance of the session.

Select materials for projects that support the therapeutic needs, keeping in mind the objectives and functional levels of the participants. Keep choices to a reasonable number so that they are not overwhelming. At the same time, allow and facilitate creativity, individuality, and choice making.

Use of space

Horticultural therapy is practiced in settings with much diversity, both in terms of physical space as well as in participant demographics, abilities, and challenges. In general terms, the following highlights areas of concern of which to be aware. Evaluate each with respect to the unique circumstances of the user group.

For some groups, it is imperative to be aware of "creature comforts" to improve participant focus. In other situations, it is important to help and encourage participants to be able to function in more challenging "natural" weather conditions in order to support therapeutic or vocational goals (Figure 5.5). When controlled environments and spaces are therapeutically desirable, attending to the following variables can minimize disruptions.

Sight and sound

- Minimize distractions from outside activity, both sound as well as visual disruptions, with the creative use of visual—and sometimes sound—barriers.

Temperature

- To the extent that the temperature can be moderated in the space, be sure it is not too hot or too cold.
- Encourage appropriate dress for the temperature of the session location, whether indoors or out.

Lighting

- Evaluate at various times of the day.

Figure 5.5 Gardening outdoors in less-than-ideal weather provides real-world experience and confidence. Photo by Rebecca Haller.

- Modify too much outdoor light or glare with awnings, freestanding or table umbrellas, and participant hats and sunglasses.
- Adjust lighting to ensure that there is enough for reading and fine motor tasks, especially important for people with visual impairment and for older adults.

Seating/workspace

- Arrange to facilitate group process
- Ensure an adequate amount of personal space as well as physical space to complete the task through the arrangement of the tables and chairs or the assignment of workspaces within a garden area.
- Arrange seating/workspace for optimal eye contact between participants and therapist.

The overall treatment space should be conducive to promoting independence. It is easier for participants to function independently in spaces that are well organized and clutter-free. Clearly labeled storage cabinets, drawers, baskets, and bins enable participants to be included in the process of gathering materials and cleaning up when appropriate.

Safety issues

In addition to being mindful of safety concerns related to the use of project materials, it is also important to be aware of how the arrangement of the treatment space impacts participant safety. Which safety concerns are relevant in each situation depends on both the characteristics of the participants and the space itself. Some issues to consider follow.

Elopement

- Be aware of entrances and exits, particularly with participants who have a tendency to wander or dash from the session.
- This is especially important during sessions where participants are not seated, such as in a garden area.

Tripping and slipping hazards

- Keep aisles and paths clear of buckets, hoses, and other equipment.
- Eliminate moss/algae growth.
- Modify flooring surface to a material that is not slippery when wet.

Access to storage

- Tools, equipment, chemicals, and other supplies that pose a hazard if misused should be stored in a locked cabinet.
- Be aware of access to the space while horticultural therapy sessions are not taking place, if the room is kept open.
- An inventory system for keeping track of locked items also alerts the therapist if an item is not returned at the end of the session (Figure 5.6).

Session review and evaluation

A thorough review of a horticultural therapy session includes examining a combination of external factors, group process, and self-assessment. Before any of these variables can be modified or improved, the therapist must realize they are present.

The review process is just as important as carefully thinking through the setup and procedures ahead of time. It also helps the therapist prepare for the next session. This procedure elevates the effectiveness of the overall program and ensures that the participants' needs are being addressed in the best possible way. There are many variables that can affect the outcome of the therapeutic experience. The therapist should review by asking questions about each component of the horticultural therapy session, including the *time, location, activity process,* and the *therapist* (see Exhibit 5.1. for session review considerations and Appendix II for a checklist).

The review process is a worthwhile endeavor because of its potential impact on improving the effectiveness of a horticultural therapy session. By answering questions like those in Exhibit 5.1, it is possible to isolate those individual elements that affect the session. With this understanding, the therapist can then more easily begin to change what is not working well and continue those aspects that are successful. The therapist may

Figure 5.6 Tools that are sharp must be carefully monitored for the safety of clients and others who may have access to them. Photo by Rebecca Haller.

create a checklist or review form that addresses the variables present in the current setting and program (see Appendix II for an example of content). It is important to note, however, that this process is completely separate from participant documentation and should not be recorded on the same form.

External factors that impact the participant experience include variables, from the time of day the session is held to the specifics of the session environment. Some of these variables may change from season to season, so do a periodic review. Additionally, there are variables that are completely out of the therapist's control, such as both positive and negative incidents and events earlier in the day or week, illness, or staff changes. Managing variables to the extent possible gives the participants the best atmosphere in which to flourish.

The session process is another opportunity to manage the group for optimal outcomes. Looking at the structure of the session as well as how the information is presented may make all the difference in how a participant is able to function in the group setting.

Exhibit 5.1   Horticultural Therapy Practice: Session Review

Begin by looking at the timing of the session. Did you have maximum attendance? Sometimes meals, medication, and transportation (internal or external) can affect participants' alertness or their ability to focus on the group session, for example. The availability of support staff to help with transportation to and from the session or to provide assistance during the session can affect success just as much as how the therapist utilizes assistance during the session. Just by considering when the session is held, a number of variables are affected.

Evaluate the space or location of the session in the same way. Look at size and whether the participants were able to maneuver comfortably. Was the seating arranged to facilitate the group process? How were the lighting and the temperature? Was it free of distractions? Consider other questions particular to your organization.

Then, take a look at what happened during the session—the activity process. Did the intended outcomes occur? Did everyone have an opportunity to participate at the maximum level possible? Maximum participation includes having ample materials and tools, appropriate modifications, and being prepared for the unexpected. It also includes having enough staff or volunteer support to enable individual participants to work on personal goals that might require standby assistance, verbal cues, or visual monitoring for safety. This is also the time to look at the timing or flow of the session. Were the participants rushed, or was there adequate time for discussion and the activity?

Perhaps the most difficult part of the review process requires the therapist to honestly look at his or her own actions, inactions, and words used during the session. For example, consider whether the instructions were presented at an appropriate level and in a logical order. Was the tone of voice used respectful rather than condescending, encouraging of participation rather than stifling? Was the best leadership style used to achieve the desired outcome? What about the group dynamics? Were you aware of and able to manage interactions among the participants? Sometimes knowing what to say to keep the session flowing is difficult. If that is challenging for you, try thinking through the introduction to a topic or the steps of a task ahead of time.

Finally, doing a thoughtful self-assessment to be sure that the best therapeutic use of one's self is employed is never time wasted (see Exhibit 5.1 for guidance).

Summary

Treatment sessions are integral to the practice of horticultural therapy. The professional process used by the therapist involves planning strategically, providing interventions, documenting outcomes, and evaluating effectiveness. Session considerations include the characteristics of the group, program type, goals and objectives, prior evidence of success, plant and human resource availability, season, garden and activity space configuration, and safety.

Bibliography

Austin, David R. 1991. *Therapeutic Recreation: Processes and Techniques*. Champaign, IL: Sagamore Publishing.

Law, Mary and Joy MacDermid. 2008. *Evidence-Based Rehabilitation: A Guide to Practice*. Thorofare, NJ: SLACK.

Stumbo, Norma J. and Carol Ann Peterson. 2004. *Therapeutic Recreation Program Design: Principles and Procedures*. 4th edn. San Francisco, CA: Pearson Education.

chapter six

Documentation
The professional process of recording treatment plans, process, and outcomes

Sarah Sieradzki

Contents

Introduction

Documentation of horticultural therapy outcomes is an essential part of professional practice. This chapter will discuss the reasons why documentation is important and offer general guidelines on ensuring its effectiveness and professionalism. Specific aspects of documentation will also be discussed, including

- Types of documentation used when providing horticultural therapy to individual clients
- Detailed information on assessments, intervention or treatment plans, progress or observation notes, and discharge summaries
- Ways to make written and verbal communication more effective
- Types of documentation used when providing horticultural therapy to groups

Purpose and importance of documentation

Whether a horticultural therapy program focuses on therapeutic, vocational, or wellness goals, documentation provides a major difference between a professional program and a nonprofessional one. The following list offers the purposes of documentation:

- Meets professional responsibility and obligations
- Provides a written record of services and treatment process
- Communicates with treatment team or other disciplines
- Monitors client progress and demonstrates outcomes
- Fulfills requirements of some regulatory or accrediting bodies

- Supplies information to obtain reimbursement for services
- Provides data for research, process monitoring, and quality assurance purposes
- Helps horticultural therapy gain credibility and recognition
- Provides information for program evaluation, adaptation, and improvement

Documentation is the written record of the entire process of assessing client function, identifying needs, establishing goals, monitoring and reporting client progress, adapting the program in response to feedback to improve client progress, and reporting positive outcomes and results of horticultural therapy. Written and verbal communication with the treatment team is essential to provide valuable information about the client's progress. Effective communication also serves to enhance the credibility and recognition of horticultural therapy as a valuable profession.

In horticultural therapy programs, documentation is an integral part of the treatment process (see Table 6.1). It creates a written medical/educational and legal record of treatment procedures and provides important information about the client to other disciplines involved. Detailed information about documenting each step of the therapeutic process will be provided later in this chapter.

In many health-care, rehabilitation, and educational settings, documentation is required by local, state, and federal laws. Knowing the specific rules of all the regulatory and accrediting bodies in each particular setting is vital to ensuring that documentation meets standards and requirements. These regulatory bodies may include the Joint Commission, Commission on Accreditation of Rehabilitation Facilities (CARF), Medicare, Medicaid, and state departments of special education, health, and mental health. Regulations are put in place to protect consumers by creating strict professional standards, to ensure quality of care, and to justify the expense of reimbursing treatment. Ethical practice requires taking the time and effort to fulfill documentation obligations. Failure to meet documentation regulations in a timely, comprehensive manner can have serious consequences, legally and financially. Insurers may also require documentation before providing reimbursement for horticultural therapy services. Attention to detail in documentation will demonstrate the quality of care being given.

In addition to the types of documentation addressed in this chapter, documentation also establishes a data record that can be collected and used for other purposes. For example, many facilities utilize process-monitoring procedures for quality assurance and performance improvement. Documentation may also serve to provide valuable information for program evaluation aimed at improving the services and outcomes of a particular horticultural program or site. Furthermore, written documentation

Table 6.1 Horticultural therapy practice: Documentation in various types of horticultural therapy programs

Type of program	Program model	Aim of program	Documentation purpose
Therapeutic	*Medical* Settings may include a hospital, skilled nursing facility, long-term care facility, rehabilitation center, pain management program, mental health setting, outpatient clinic, group home, special education setting, substance abuse/ addictions recovery program, or hospice.	Management or recovery from an illness or injury; improved quality of life and functional independence.	Demonstrate how horticultural therapy activities help clients meet treatment goals; provide information to other disciplines or to the treatment team; meet requirements of regulatory bodies.
Vocational	*Habilitation or rehabilitation* Settings may include special education settings, programs for people with intellectual or developmental disabilities, sheltered workshop, rehabilitation center, work hardening programs, correctional facility (prison), programs for youths at risk, traumatic brain injury programs.	Increase specific vocational skills to maximize employability; maximize functional independence and ability to work with others.	Demonstrate how horticultural therapy activities help clients meet vocational training goals; provide information to other disciplines; meet requirements of regulatory bodies.
Social and wellness	*Wellness* Settings may include community gardens, arboreta, older adult day care, community wellness programs, garden clubs, mental health day treatment programs, social service agencies, support groups, and dementia programs.	Enhance personal satisfaction, quality of life, sense of well-being; increase ability to improve wellness, prevent illness, or cope with health issues.	Demonstrate how horticultural therapy activities can enhance the wellness of individuals/ groups and improve quality of life.

data can be collected, measured, described, and reported for research purposes to increase the body of evidence regarding the value and benefits of horticultural therapy. Outcome studies such as these improve the professional image and visibility of the profession.

General guidelines for horticultural therapy documentation

Because of differing target audiences, populations, regulations, and types of horticultural therapy programs, each setting is required to tailor its documentation to meet specific needs. To maximize its effectiveness, however, all documentation benefits from following certain guidelines as described in the following list:

- Timely
- Signed and dated
- Organized and easy to read
- Site or population specific
- Clear and understandable
- Legible and neat
- Professional and well written
- Meaningful and relevant
- Accurate and specific
- Descriptive and objective
- Kept confidential
- Directed at the target audience

These suggestions are also helpful in the process of designing the particular format of documentation to be used in an individual horticultural therapy setting.

No matter what recording method is selected, documentation must be completed and shared in a timely manner for it to be meaningful to the target audience. In health-care and educational settings, there may be regulations that set time parameters for when different types of documentation must be completed. It is important to know the unique requirements of each horticultural therapy setting so that documentation can be completed by the required deadlines. It is easy to become caught up in the more interesting and creative tasks of program development and in providing horticultural therapy to a given population and not leave adequate time for essential documentation. It is critical to build documentation time into the daily schedule so it can be completed in a timely manner.

Keep in mind that documentation creates a legal record of treatment provided, so it is important to sign and date each document. When signing a document, place professional credentials after the signature.

Keep documents well organized and easy to read. It is generally helpful to use as simple a format as possible. When writing, be direct, concise, and to the point. Include only important observations and pertinent details and keep related points together. Add emphasis to the most significant points by stating them first and then follow through in order of importance.

Use a format that is site or population specific and determine content based on the needs of target audiences. When developing the format for documentation, it is crucial to meet the needs of the target audience that will read the notes. This may include regulatory and accrediting bodies, insurers that are asked to reimburse services, other disciplines that form the treatment team, family members, and in certain instances, the client. The format can be designed as simply or as comprehensively as required for the needs of the program. It is also helpful to consider the time that will be allotted for documentation purposes. Checklists, charts, surveys, or graphs may be ways to simplify writing tasks rather than using more time-consuming narrative notes. Networking with other horticultural therapists in similar settings may provide ideas for easy and effective documentation formats.

Strive for clarity when documenting horticultural therapy procedures and client progress. Avoid unnecessary jargon or terms that could be easily misunderstood. Do not use confusing abbreviations unless a key is readily available on the document. Ambiguous terms such as "fair" or "poor" are usually not helpful as a measurable description.

Any handwriting included should be legible and neat. Keep in mind that the appearance of a document says a lot about the time and care taken to create it. High-quality documentation is a tangible representation of a quality horticultural therapy program. Black pen is generally considered best for professional documentation as it is most readable. If a writing error occurs, the legal way to correct it is to cross it out with one line and add the notation "Error" along with the writer's initials.

Documentation should be professional and well written. Always use correct spelling and punctuation when recording horticultural therapy documentation. When writing narrative notes, use complete sentences, correct sentence structure, and good grammar. Documentation is usually written in the third person—using "staff" or "therapist" instead of "I" or "me."

Write to express clear ideas rather than to impress someone with jargon or flowery language.

Make documentation meaningful and relevant. Document the most important information that will be useful to the target audience. This usually means focusing on the client's functional abilities and progress toward established treatment goals. Horticultural therapy has a unique perspective in observing clients actually performing an activity. Thus, the therapist may have pertinent information to communicate regarding the task performance and organizational skills of the client.

Written or verbal communication should be accurate and specific. It is helpful to be as specific as possible when documenting. Vague reporting will cloud the significance of client and activity observations. Wherever possible, use measurable terms and record factual information. Bias has no place in professional documentation. It is crucial to be truthful, even if the horticultural therapy procedures were not as satisfactory as desired, or when goals remain unachieved.

Be descriptive and objective in reporting. It is usually best to be factual in reporting observations of what actually occurred during horticultural therapy sessions. Be careful not to make assumptions or use inferences when reporting—staff commentary that indicates bias or personal opinion is inappropriate. Value judgments such as "nice," "pleasant," and "well behaved" are also not useful or appropriate. It is helpful to use measurable terms (see Appendix I for examples) whenever possible and to use descriptive language to tell the story of what occurred during horticultural therapy sessions. Write as holistically and comprehensively as possible, given the time and space restraints. Do not use ambiguous terms that are open to misinterpretation.

All documentation about clients should be kept confidential. It should be shared only with necessary treatment team or other staff members. If documentation will be shared with insurers or other agencies, permission must be obtained from the client or the client's legal guardian. Many local, state, and federal regulators have strict confidentiality standards that must be followed; failure to do so may have legal consequences. (See more detailed information about confidentiality requirements in Exhibit 6.1.)

Individualized documentation

Considerations

Therapeutic and vocational types of horticultural therapy programs generally require documentation to be completed regarding each individual client following every step in the therapeutic process (see Table 6.2). Some documentation may be mandated, such as an individual education plan (IEP) in a special education setting. Some settings may have set formats for documentation; in other settings, horticultural therapy staff may be responsible for designing their own forms for documentation. Some settings may utilize computer-generated documentation, while others will rely on written and/or verbal reporting.

Types of individualized documentation

Each step in the therapeutic process may require a different type of document to capture the necessary information for reporting horticultural

**Exhibit 6.1   Horticultural Therapy Practice:
Confidentiality Requirements and
Electronic Medical Records**

All documentation by horticultural therapists in health-care settings is subject to federal privacy and confidentiality laws. This would include horticultural therapists who work in hospitals, skilled nursing facilities, extended care facilities, assisted-living facilities, mental health settings, and hospices, and therapists working for private medical practices. All other horticultural therapy settings and areas of practice may be subject to differing rules and regulations. Therapists in settings that are *not* health care or medical in nature should identify the applicable state or federal rules regarding their documentation.

The U.S. Department of Health and Human Services (HHS) has established specific confidentiality standards that must be followed to protect the private health information of all clients in health-care settings. These standards were issued to help health-care agencies implement the requirements of the Health Insurance Portability and Accountability Act of 1996 (known as HIPAA). This law was designed to enable individuals to maintain their health insurance, to protect private health information securely and confidentially, and to increase the efficiency of health-care organizations so they can better control their administrative costs. The HHS privacy standards require that all "individually identifiable health information" stored or transmitted by health-care organizations as well as their business associates be securely protected, no matter what form or media is used—electronic, paper, or verbal. Therapists who work for a health-care organization, or private contractors who would be considered business associates, must comply with these standards. The penalties for disclosure of protected individually identifiable health information are dire and can include both monetary and criminal penalties. Protected information includes the client's name, address, telephone number, date of birth, and social security number. Demographic information and diagnoses relating to a person's past, present, or future physical or mental health conditions are also protected.

Additionally, the federal government is promoting the development and use of electronic medical records (EMR) to improve the safety, quality, privacy, and efficiency of the provision of health care in the United States. EMR are designed to improve patient access to their own health records and to health information resources to improve their ability to manage health conditions. EMR may also make it easier for physicians and other practitioners to access laboratory results and test results from multiple sources and settings, and to order medical tests and medications. EMR can assist therapists in scheduling treatment sessions and provide secure ways to convey communication about clients among other treatment team members.

Therapists who use electronic or digital methods of collecting and storing protected health information must ensure that these records cannot be accessed, even if their digital device is lost or stolen. Most health-care organizations will provide staff with laptops, tablets, or telephones that encrypt all protected information. If a therapist plans to use his or her own device for storing or transmitting personally identifiable protected information, he or she needs to protect this information with at least firewalls and very secure passwords. Encryption is the ideal for transmitting and storing protected health information and personally identifiable demographic information.

The HIPAA regulations do allow the transmittal of some protected private health information to other government agencies in a few specific instances. Examples of these exceptions include national security needs, information to ensure public health and safety in emergencies (such as an outbreak of communicable diseases such as measles, HIV, or Ebola), and protection of vulnerable individuals such as victims of sexual abuse, domestic violence, and elder abuse, or others who lack the capacity to protect themselves.

Health and demographic personal protected health information is sometimes sought for use in research purposes. Practitioners must make a formal research proposal to their organization's institutional review board. These boards regulate the use of protected health information and provide safety and confidentiality guidelines for research purposes. Totally de-identified health information (information without any possible identifying data such as name, age, sex, address, phone number, diagnosis, or other protected demographic and health data) is most easily approved for research purposes. The institutional review board will set specific parameters and protections for all human research—these must be followed carefully and will be monitored by the review board throughout the research process. Researchers may need to demonstrate their understanding of the confidentiality laws and safety protections as part of their research proposal.

REFERENCES

U.S. Dept. of Health & Human Services. *Summary of the HIPAA Privacy Rule.* May 2003. http://www.hhs.gov/hipaa/for-professionals/privacy/laws-regulations/ (accessed March 20, 2015).

U.S. Dept. of Health & Human Services. *Electronic Medical Record Systems.* February 2015. http://healthit.ahrq.gov/key-topics/electronic-medical-record-systems (accessed March 20, 2015).

Table 6.2 Horticultural therapy practice: Documentation and the therapeutic process

Steps in the therapeutic process	Documentation that may be applicable
Initial assessment to determine client's abilities and identify needs	Initial evaluation report
Establishment of treatment goals and objectives as part of the intervention plan	Written treatment plan that includes long-term goals, short-term objectives, how often specific interventions will be used, target dates for goal achievement, and who is responsible for providing treatment
Intervention through horticultural therapy activities (individual or group setting)	Group or individual treatment session observation notes, including record of adapted tools, techniques, and methods used
Monitoring and reporting progress through ongoing reassessment	Progress notes, including goal updates, at prescribed intervals
Modification of treatment to maximize progress	Adaptation or addition of goals and progress notes providing record of modifications and resulting progress
Reporting outcomes/results of treatment	Discharge summary, including goal outcomes

therapy process and outcomes. Each type of document is described in detail in the following chapter section, including the document's purpose, methods and procedures, timing, possible formats, and content areas. (Examples of documents are found in Appendix III.)

Initial assessments

Purpose

An initial assessment is conducted to determine the present level of functioning of the client and to identify treatment needs. The specific format, method, and content of the initial assessment are dependent on the site and population of the facility as well as on requirements of regulatory and accrediting bodies.

Methods and procedures

There are a variety of ways to determine the present level of function of the client and needs for treatment, including self-reporting surveys, performing standardized evaluations, or using observations of the client performing evaluative tasks. A situational assessment is the observation of a client's behavior during a horticultural activity. A work sample is a standardized, highly structured, concrete task completed within a specific time frame. Measurements are taken regarding the client's accuracy and the amount of work completed during the timed task.

Timing

The initial assessment is completed as soon as possible after the client is referred and prior to the development of treatment goals and the intervention plan. Some settings have mandated time frames for completion of the evaluation process.

Possible formats

Initial assessments can take various formats from simple checklist self-surveys to more comprehensive charts. In some settings where ongoing reassessments are done to determine the client's progress, the initial assessment form might include additional columns in which to document these periodic updates (see Appendix III for an example of an assessment format).

Content areas

Most initial assessments identify a client's strengths and abilities as well as limitations and needs. Specific content varies, depending on the requirements of the site, population, regulatory bodies, and target audiences. In vocational settings, most initial assessments include information about a client's cognitive function, task performance skills, cooperative attitude and behavior, and prevocational skills. In health-care settings, assessment content may include physical neuromuscular function such as fine and gross motor skills, range of motion, muscle strength, endurance, and sensation. It may also include detailed information about cognitive function, task performance skills, communication skills, leisure skills, and interpersonal skills.

Intervention plan or treatment plan

Purpose

A treatment or intervention plan outlines the specific way that horticultural therapy staff (and other involved disciplines) meet the identified needs of the client. When done well, it provides a road map for the interventions that will follow and sets parameters for goal achievement and eventual discharge from treatment (see Appendix III).

Methods and procedures

In some health-care and educational settings, a multidisciplinary treatment plan is mandated, which includes timing, the basic format, and content. The entire team may work together to develop the intervention plan. For example, in special education settings, state or federal laws may set requirements for an IEP (as shown in Exhibit 6.2). In an acute-care hospital setting, the Joint Commission may require specific timing, contents, and procedures for a client's interdisciplinary treatment plan (ITP). In a rehabilitation setting, CARF may set the standards for the plan of care. In some settings, the horticultural therapist may develop the treatment plan individually. Wherever possible, the client is included in the identification and establishment of goals. This will lead to client-centered practice in which the top priority goals are most meaningful to the client, which results in increased motivation during treatment. (See Table 2.1 in Chapter 2 for information about who is typically included on treatment teams.)

Timing

The intervention or treatment plan is written as soon as possible following the initial assessment and prior to the start of actual therapy sessions.

Possible formats

The treatment plan format may be mandated by a regulatory or accrediting agency (see "Methods and procedures" in this chapter) or developed at the site of horticultural therapy practice. In some settings, the treatment plan is an integral part of the initial assessment format, while at other sites it is an independent document.

Content areas

The intervention plan generally includes the client's identified needs or problem areas, long-term goals, short-term behavioral objectives, specific interventions to be used (group or individual treatment sessions), how often and how long treatment will occur, which specific disciplines or persons are responsible for carrying out treatment procedures, and target dates for goal achievement. Creating effective treatment goals and objectives is key to easy measurement of outcomes later in the process.

A simple technique for writing effective goals is to use the SMART method outlined here. SMART goals are

- Specific: state exactly what the client will accomplish
- Measurable: have the number of items or times something will be done

**Exhibit 6.2   Horticultural Therapy Practice:
Components of an Individual Education Plan**

The federal government has mandated some specific components
for IEPs for school-age students with an identified disability. State
and local boards of education may require additional content for
IEPs as well. In general, the customary IEP components are

- A description of the child's present levels of educational and/
  or functional performance, including academic achievement,
  prevocational and vocational skills, psychomotor skills, and
  self-help skills
- An indication of how the child's disability impacts his or her
  educational performance
- Annual goals that describe what the child can reasonably be
  expected to achieve during the current school year and specific
  criteria for measuring goal accomplishment
- Short-term objectives that provide measurable steps toward
  achievement of the annual goals and how the parents will be
  regularly informed of the child's progress
- Identification of (and justification for) the particular special
  educational and related services needed by the child, along
  with any specific accommodations (such as aids, modifications,
  and supports) required in the classroom or during standardized testing to meet that child's unique needs
- The extent to which the child will participate in regular education programs with typical children to help ensure the least
  restrictive placement for the child
- Specific information regarding services to be provided, including beginning and ending dates, frequency, location, and duration of services, and who will be responsible for providing them
- For children aged 14 and older, identification of transitional
  services (such as linkage to other agencies and interagency
  responsibilities) needed to help the child reach post-school
  goals such as post-secondary education, vocational training,
  independent living, community participation, and employment (including supported employment)

- Action oriented: tell how the goal will be achieved—stated using a
  verb
- Realistic: enough to be challenging, but not too difficult
- Time based: have an end point or deadline

An effective goal is specific, measurable, action oriented, realistic, and time based (has a deadline). A SMART goal succinctly states what the client will accomplish in a particular way by a certain deadline. SMART goals can be written for both individuals and groups who receive horticultural therapy services. At the conclusion of the therapy session(s), the therapist and client can quickly and easily assess whether the goal has been accomplished, and documentation of progress will be simpler.

Knowing what to measure is often the most difficult part of goal writing and documentation. (See Appendix I for a broad array of horticultural therapy short-term goals, examples of treatment activities, and what to measure to show treatment effectiveness in various types of horticultural therapy programs.) Long-term goals are often to improve functional abilities, independence, or quality of life of the client.

Progress notes

Purpose

Progress notes are the ongoing record of horticultural therapy treatment sessions and changes in the client's functional status in response to this treatment.

Methods and procedures

In some horticultural therapy programs, observation notes may be written following each individual or group therapy session. In other settings, progress notes are written periodically, often on a weekly basis.

Progress can be measured in either quantitative or qualitative terms. When using a quantitative method, progress is expressed in a countable, comparable, quantifiable way—usually using a checklist, graph, or chart of duration, percentage of accuracy, or number of completed tasks. When using a qualitative method, the progress note demonstrates changes in client behavior or function through a descriptive characterization such as a narrative or case study.

There are several ways to document progress: a change in the level of assistance the client needs to perform a specific task, a change in response to treatment (example—increased consistency of attendance), development of a new skill or ability, or learning a new compensatory technique.

It is also important to document temporary setbacks such as slow (or complete lack of) progress due to pain, medication change, illness, or medical complications, or external circumstances such as family or scheduling issues. When setbacks are identified, it is important to state the perceived reason and the plan of action to enhance future progress.

Timing

In some settings, the frequency of progress notes is mandated, while in others, frequency will be determined by time and space constraints. A progress note is written weekly in many settings. For long-term settings such as rehabilitation centers, sheltered employment, or residential programs, the interval between progress notes may be longer.

Possible formats

Documentation of progress may take the form of checklists, rating scales, or narrative notes. If the facility uses a problem-oriented medical record, the format used by all disciplines may be the SOAP format described here. SOAP documentation includes the following:

- Subjective information from the client
- Objective observations of staff and clinical findings
- Assessment of progress toward established treatment goals or interpretation of reassessment data
- Plan for further treatment including any potential changes to enhance progress

In other settings, graphs, surveys, or charts may be used. For example, in vocational settings, documentation may include detailed information about performance on the steps of a horticulture task or demonstration of positive social behaviors. When selecting or developing a format, emphasis should be placed on including the most comprehensive amount of information in the simplest and most concise fashion to expedite documentation duties. (See Appendix III for an example of a form used for recording client progress.)

Content areas

Progress notes will often include a record of attendance, the level or quality of participation, a description of treatment sessions, changes in client behavior or function, other relevant observations, the therapist's interpretation of the reassessment data, an update on progress toward goal achievement, and the plan for further horticultural therapy involvement. The future plan may include the modification of goals or the adaptation of horticultural therapy procedures to enhance client progress, or the identification of new goals if the first objectives have been accomplished.

Discharge summary

Purpose

The discharge summary is a report that recaps the entire treatment process and records the final reassessment and update on goal achievement.

It is generally a brief synopsis of outcomes; however, it needs to be as comprehensive as possible within a concise and simple format. Discharge summaries are often the document most read by insurers to determine if services will be reimbursed and by regulators to ensure quality of care.

Methods and procedures

Some facilities use a separate format for a discharge summary, while others write a more detailed final progress note. Most discharge summaries document the amount of goal achievement shown as well as recommendations for follow-up care.

Timing

The discharge note should be written as soon as possible following the discontinuation of horticultural therapy services. In some settings, regulators mandate the time frame.

Possible formats

The discharge summary document can be a narrative note, checklist of progress, self-evaluation such as a satisfaction survey, or graphic representation of results such as a chart or graph of progress.

Content areas

The discharge summary generally contains information about attendance, the number of sessions the client experienced, a reassessment of the client's functional abilities and an update of goal achievement, a review of any adaptive tools/methods/techniques used, any needed interpretation of results, and recommendations for further follow-up care. (See Appendix III for an example of discharge summary contents and format.)

Verbal reports

Purpose

Horticultural therapists are often called upon to give verbal reports of client progress in team meetings and conferences. Similar principles of good communication apply to verbal reporting as to written notes.

Methods and procedures

If possible, prepare ahead of the meeting for what needs to be included in the verbal report. Just as in written records, emphasize the most significant information by reporting it first and then follow in descending order of importance. Be succinct and only report pertinent and meaningful observations. Many team conferences are held as a forum, combining information from numerous sources to determine the future course of treatment for a client or for group problem-solving purposes.

Timing

Conferences may be held at any time throughout the treatment process; however, in certain settings, team meetings may have a mandated frequency.

Possible formats

Verbal reports may be given as part of an IEP meeting in an educational or vocational setting or as part of an ITP meeting in a medical or rehabilitation setting. Meetings may also be held with administrative staff, the client, family members, or a case manager from another involved social service agency. In community horticulture programs that do not necessarily include a full treatment planning process, verbal reports may be the primary means used to communicate the progress and development of individual participants.

Content areas

Often, the individual responsible for scheduling the conference provides an idea of what information or problem will be discussed. This allows the horticultural therapist to prepare specific examples of goal progress or pertinent adaptive methods to share with others. Having the opportunity to prepare ahead of time may even allow for the development of graphic or written supplementary materials to provide more comprehensive information to the group. Information content may vary, but usually will include details regarding client response, behaviors, problem areas, and progress toward goals.

Group documentation

Because horticultural therapy is often provided to groups of clients, some programs may utilize group documentation in addition to, or in place of, more highly detailed, individualized documentation. Group documentation will usually be employed when the program uses group goals rather than individualized objectives for each client. Group documentation may also be used when the horticultural therapy group meets infrequently—monthly, for example. Although documentation about individual clients within a group may also be recorded, the methodology needs to be simpler and less time-consuming than that described in the previous section when recording about a larger number of clients.

In this section, group documentation will be described, including its purpose, methods and procedures, timing, possible formats, and content areas.

Purpose

Social or wellness types of horticultural therapy programs often involve providing activities to meet group goals. Group documentation will record the needs, goals, and outcomes of group horticultural therapy programming.

Methods and procedures

The documentation of group horticultural therapy programs follows a similar pattern of steps in the therapeutic process as just described for individualized documentation. An assessment procedure takes place prior to the initiation of horticultural therapy sessions to determine the needs and goals of group members. A report of observations of each group session may be desirable if time allows. Reviewing this information assists the horticultural therapist in assessing ongoing progress and client satisfaction. The therapist can then make any needed activity modifications to enhance the program for group members. At the conclusion of the predetermined number of group sessions, a final summary report outlines goal outcomes and any other pertinent information about the horticultural therapy experiences of the group.

Timing

When programs are offered with a beginning and an end, assessment and goal development should take place in the first group session and reassessment should occur in the last horticultural therapy session of the group program. For ongoing groups, periodic documentation is performed.

Possible formats

Simpler versions of individual client documentation may be utilized to collect and record group session information. Using a client survey as a pre-assessment, and then again as a post-assessment, will provide a comparative method of self-evaluation. The use of rating scales with results that can be compared visually using a chart or graph is another easy method of documentation. (See Appendix III for an example of a group documentation format.)

Content areas

Contents of group documentation may be similar to those of individualized documentation or they may be entirely different, depending on the needs and goals of the group program. Social and wellness types of horticultural therapy programs may emphasize personal satisfaction and quality-of-life issues rather than specific skill development or functional capability. (See Appendix I for sample SMART goals, treatment activities, and ideas for measuring progress for social and wellness types of horticultural therapy programs.)

Summary

This chapter described the customary methods and content of individual and group client documentation used in professional horticultural therapy practice. Performing documentation effectively is crucial for the credibility and future growth of the profession.

Bibliography

Austin, David R. 1991. *Therapeutic Recreation: Processes and Techniques.* Champaign, IL: Sagamore Publishing.

Best-Martini, Elizabeth, Mary Anne Weeks, and Priscilla Wirth. 2011. *Long Term Care for Activity Professionals, Social Services Professionals, and Recreational Therapists.* 6th edn. Enumclaw, WA: Idyll Arbor.

Borcherding, Sherry. 2000. *Documentation Manual for Writing SOAP Notes in Occupational Therapy.* Thorofare, NJ: SLACK.

Davis, William B., Katie E. Gfeller, and Michael H. Thaut. 1992. *An Introduction to Music Therapy: Theory and Practice.* Dubuque, IA: Wm. C. Brown.

Ozer, Mark, Otto D. Payton, and Craig E. Nelson. 2000. *Treatment Planning for Rehabilitation: A Patient-Centered Approach.* New York: McGraw-Hill.

Simson, Sharon P. and Martha C. Straus. eds. 1998. *Horticulture As Therapy: Principles and Practice.* New York: The Haworth Press.

Stumbo, Norma J. and Carol Ann Peterson. 2004. *Therapeutic Recreation Program Design: Principles and Procedures.* 4th edn. San Francisco, CA: Pearson Education.

Appendix I: Goals, activities, and measurements

SMART goals, activities, and measurement for the three types of HT programs

The following are examples of SMART goals (also referred to as "objectives") that might be applied in various horticultural therapy (HT) settings. For each type of program—vocational, wellness, and therapeutic—potential settings and diagnoses are specified, with ideas for goals/objectives, activities, and what to measure.

Vocational HT programs

Vocational training or rehabilitation

- *Settings:* Work hardening clinics, sheltered workshop for intellectual disability/developmental disability (ID/DD) clients, traumatic brain injury (TBI) program
- *Typical Diagnoses and Other Types of Participants:* DD, ID, youth at risk, TBI, correctional facility (prison) population, hand injuries, back injuries, special education, spinal cord injuries

Short-term goal/objective	HT activity	What to measure
Client will remember and follow three-step directions with 80% accuracy without visual or verbal cues after two treatment sessions.	Plant stem tip cuttings to be grown for sale.	Compare trials of same three-step directions over time to determine accuracy percentage.
Client will add prices of sale plants and give correct change with 80% accuracy.	Perform cashier job at plant sale.	Determine accuracy percentage of several trials.

(Continued)

Short-term goal/objective	HT activity	What to measure
Client will demonstrate accurate counting skills from 1 to 12, 100% of the time after five learning trials.	Each trial will consist of placing 12 pots in flat to be used for potting up plant divisions.	Determine accuracy of 12 pots per flat in several trials.
Client will clean worksite after task completion with no verbal reminder after three introductory learning trials.	Use whisk broom and dustpan to clean up potting mix from table and floor and then wash hands with soap and water.	Chart number of times per week that client proceeds with appropriate cleanup tasks on conclusion of work session without verbal reminder.
Client will be on time and participate fully in the entire HT activity for 9 out of 10 sessions.	Develop a community-based vegetable garden in the vacant lot near the community youth center.	Keep track of client's active participation and tardiness.
Client will follow the expectations for positive social skills formulated by the group for 10 consecutive sessions.	Discuss the needs of seeds for optimum growth and relate these to the need for identified specific positive social behaviors for individual growth.	As a group, create a short list of expectations for positive social behavior, communication, and cooperation. Have participants track their positive behaviors after each program session.

Social and wellness HT programs

Community wellness programs

- *Settings:* Botanical garden or arboretum, community health center, support group, nonprofit health agency (such as National Multiple Sclerosis Society, American Heart Association, American Cancer Society, Alzheimer's Association, National Alliance for the Mentally Ill), group homes, day treatment programs, long-term care residences
- *Typical Diagnoses and Other Types of Participants:* Individuals with multiple sclerosis (MS), heart disease, cancer, arthritis, or other chronic or debilitating illness; caregivers of individuals with chronic illness or disability; individuals who are dealing with grief and loss, elders, older veterans

Short-term goal/ objective	HT activity	What to measure
During first support group session, client will set two SMART goals to positively impact physical or emotional health.	Pot cuttings of plant that best represents how client would like health to be in the future (selected from a variety of plants). HT activity leads to small-group discussion of SMART goals.	Note client's written SMART goals and action plan of ways to positively impact health.
Client will identify two positive ways to better cope with illness and stressors by the end of three support group sessions.	Create potpourri using dried flowers, dried fragrant herbs, and spices. Activity leads to a discussion of the use of fragrance for relaxation and other methods of self-care.	Note written or verbal identification of two potentially helpful coping strategies.
Client will discuss a way to better cope with grief and loss by the conclusion of two support group sessions.	Arrange pressed flowers to be put in a frame, using Victorian meanings of flowers and leaves to describe loved one. Discussion of grief and loss will accompany activity.	Note written or verbal identification of a way to safely express feelings and better manage emotions as part of grief process.

Therapeutic HT programs

Physical rehabilitation

- *Settings:* Rehabilitation center, skilled nursing facility, home health care, orthopedic/sports medicine program, work hardening program
- *Typical Diagnoses:* Cerebrovascular accident (CVA), hip fractures, orthopedic/sports injuries, burns, TBI, spinal cord injuries, MS, neurological disorders, hand injuries, repetitive use injuries such as carpal tunnel or tennis elbow, joint-replacement surgeries

Short-term goal/objective	HT activity	What to measure
Client will extend horizontal reach by 10 inches forward and 6 inches to each side after three treatment sessions.	Reach forward and side-to-side to plant a tray of seeds.	Record length and arc of forward reach in inches.

(Continued)

Short-term goal/objective	HT activity	What to measure
Client will tolerate standing for 30 min after two sessions.	Pot up plant cuttings while at standing table.	Record length of standing tolerance time from session to session.
Client will improve muscle strength from poor to fair and length of endurance for upper extremity tasks up to 20 min without a break after five treatment sessions.	Fill pots with soilless mix as part of group assembly line potting task.	Use muscle testing or record increase in weight of pots with soil. Record the length of time client works on task before tiring and needing a break.
Client will improve pincer grasp by increasing number of seeds sown by 100% after three treatment sessions.	Plant seeds of various sizes by using pincer grasp.	Record speed and accuracy of picking up and planting seeds (comparison of timed trials).
Client will show improved eye–hand coordination by improving speed and accuracy of watering task by 50% over three treatment sessions.	Water trays of small pots using squirt bottle or watering bulb.	Record speed and accuracy of watering task (comparison of several timed trials).

Pediatric health care

- *Settings:* Acute pediatric hospital unit, special child day treatment program, school-based special education setting, pediatric rehabilitation center
- *Typical Diagnoses:* Cerebral palsy, autism, mental retardation (ID)/ DD, burns, fractures, muscular dystrophy, accident trauma, poisonings, brain injuries, surgical procedures, childhood cancers

Short-term goal/ objective	HT activity	What to measure
Client (receiving chemotherapy) will explore at least two ways to cope with impending hair loss to reduce impact of feeling different.	Create "grass goon" creature with facial features to try out various grass "hairstyles" leading to small-group discussion of issue.	Note identification of two ways to cope with embarrassment of impending hair loss following chemotherapy treatment.

Short-term goal/ objective	HT activity	What to measure
Client will demonstrate improved eye–hand coordination and pincer grasp by increasing accuracy of seed sorting by 80% in three treatment sessions.	Sort seeds (of a variety of sizes and shapes) into small containers, then choose seeds to plant in own peat pots to take home to be planted.	Measure timed trials for accuracy of seed sorting.
Client will identify two ways to contribute to own recovery and maintain improved health/wellness after discharge in three treatment sessions.	Adopt a plant that needs frequent care to experience nurturing another living thing (rather than always being recipient of care). Participate in discussion of best ways to care for plant and self in the future.	Note identification of two ways to better care for own health/wellness after discharge from hospital setting.
Client will cooperatively practice gait training in two physical therapy sessions (working cooperatively with horticultural therapist to co-treat client).	Perform planting activity with materials in various places around the room.	Note walking around room to obtain various supplies for activity with few or no complaints.

Mental health

- *Settings:* Inpatient acute-care behavioral health unit, community mental health center, partial hospitalization program, group home, home health, addiction recovery, homeless shelters, veterans hospitals, disaster shelters
- *Typical Diagnoses:* Major depressive disorder, bipolar disorder, anxiety and panic disorders, schizophrenia, personality disorders, post-traumatic stress disorder

Short-term goal/ objective	HT activity	What to measure
Client will explore one or two healthy and productive leisure pursuits for use post-discharge during two to three treatment sessions.	Explore houseplant varieties and care, as well as the development of a small container garden to take home.	Note identification of one to two leisure pursuits to be used at home post-discharge with detailed information on what, when, how, and where.

(Continued)

Short-term goal/ objective	HT activity	What to measure
Client will demonstrate cooperation and positive social skills in three treatment groups prior to discharge.	Perform horticultural tasks such as seed sowing and planting stem tip cuttings to take home.	Note cooperation with a group of peers without disruption or agitation in three group sessions.
Client will demonstrate improved communication skills by verbally requesting materials and seeking help to clarify directions in two treatment sessions prior to discharge.	Plant an edible flower salad container garden or make potpourri (tasks that involve use of a variety of materials and have multiple-step directions).	Note asking for materials and verbally requesting assistance or clarification of instructions in two treatment sessions.
Client will identify two or three positive coping strategies to better manage illness and stressors post discharge.	Perform plant division activity in small group leading to discussion of meeting needs of plant and self to promote positive growth.	Note identification of two or three methods of coping to better care for meeting own needs post discharge.
Client will demonstrate adequate behavioral self-control to complete one 30-min horticultural task without agitation, aggressive behavior, or verbal outburst.	Perform horticultural tasks such as grooming dead leaves off scented geranium plants, and potting up offshoots from succulent plants.	Note tolerating 30-min tasks including staff instructions/redirection without angry verbal outburst or agitated behavior.
Participant will verbally identify and practice at least one gardening activity that may be self-soothing during three consecutive HT sessions.	Use rhythmic garden activities that include the use of rocking motions, such as raking, hoeing, cultivating, and weeding, to learn more effective ways to cope with anxiety and panic.	Record verbal identification and participation.

Pain management

- *Settings:* Pain management clinic, home health care, rehabilitation center, skilled nursing center, orthopedic/sports medicine program, and hospice

- *Typical Diagnoses:* Arthritis, back injuries, shingles, cancer, post-surgical pain, fractures, repetitive use injuries, migraines, degenerative spinal discs, burns

Short-term goal/ objective	HT activity	What to measure
Client will demonstrate reduced intensity of pain perception during HT activity by two points on pain scale after two treatment sessions.	Pot up scented geranium cuttings or cut roses and separate petals to dry for potpourri-making purposes.	Record change in perceived pain intensity using 1–10 pain scale (comparison of trials).
Client will identify two new coping strategies to better manage pain after two treatment sessions.	Plant an edible flower salad garden in a large container in small group.	Note identification of two new pain management strategies.

Hospice

- *Settings:* Residential hospice unit, home health care
- *Typical Diagnoses:* Terminal cancer, congestive heart failure, and degenerative neurological disorders

Short-term goal/ objective	HT activity	What to measure
Client will experience reduced perception of pain by two points on pain scale during and immediately following a HT treatment session.	Arrange bouquet of cut flowers to be placed in client's room or given to loved one.	Record change in perceived pain intensity of two points on 1–10 pain scale during HT activity and for 30 min afterward
Client will express sense of peace and closure regarding impending death in one HT treatment session.	Discuss type of tree to be planted to leave legacy for loved ones and discuss end-of-life issues.	Record verbal expression of readiness for natural death or sense of peace in having closure with loved ones.

Specialized dementia

- *Setting:* Alzheimer's unit in skilled nursing facility, geriatric adult day care program, home health care, inpatient geropsychiatry unit

- *Typical Diagnoses:* Alzheimer's dementia, multi-infarct dementia, Pick's disease, Huntington's disease, Crutchfield-Jacob's disease, traumatic brain injury, CVA

Short-term goal/ objective	HT activity	What to measure
Client will demonstrate orientation to self during HT treatment activity after three treatment sessions	Care for houseplants brought to unit on cart along with reminiscence discussion of plants, gardening.	Record verbal expression of a memory of past gardening experiences or identification two types of plants.
Client will demonstrate reduced agitation and ability to socialize appropriately with peers for 15 min during HT treatment activity.	Divide and plant fragrant herbs into container gardens for unit courtyard or plant heirloom plants in containers.	Record that client verbally addresses staff or peers in appropriate manner for 15 min during activity. (Client will display no disruptive behavior.)
Client will follow two-step directions with verbal cues for 15 min in three HT treatment sessions.	Arrange bouquet of fresh or dried flowers to place in client's room (nontoxic plants only).	Record following two-step directions for 15 min.

Source: (Contributed by S. Sieradzki)

Appendix II: Activity, task, and session planning resources

Included in this appendix is a sampling of resources to help the therapist plan, schedule, organize, and review the horticulture tasks and activities as well as the process and structure of horticultural therapy sessions.

Holidays and significant celebrations

- *January (Flower: Carnation)*
  - New Year's Day
  - Martin Luther King Jr. Day
  - Chinese New Year
- *February (Flower: Violet or Sweetheart Rose)*
  - Groundhog Day
  - Lincoln's Birthday
  - Valentine's Day
  - President's Day
  - Washington's Birthday
  - Mardi Gras
  - Ash Wednesday
- *March (Flower: Jonquil or Daffodil)*
  - St. Patrick's Day
  - Spring Solstice
  - National Horticultural Therapy Week
- *April (Flower: Daisy or Sweet Pea)*
  - April Fool's Day
  - Palm Sunday (sometimes in March)
  - Passover
  - Easter (sometimes in March)
  - Earth Day
  - Arbor Day (last Friday of April)

- *May (Flower: Lily of the Valley)*
  - May Day
  - Mother's Day
  - Armed Forces Day
  - Memorial Day
- *June (Flower: Rose)*
  - Flag Day
  - Father's Day
  - Summer Solstice
- *July (Flower: Cornflower)*
  - Independence Day
- *August (Flower: Gladiolus)*
- *September (Flower: Aster)*
  - Labor Day
  - Patriot Day
  - Grandparent's Day
  - Rosh Hashanah
  - Fall Solstice
  - Yom Kippur
- *October (Flower: Calendula)*
  - National Children's Day
  - Columbus Day
  - United Nations Day
  - Halloween
- *November (Flower: Chrysanthemum)*
  - Election Day
  - Veteran's Day
  - Thanksgiving
- *December (Flower: Poinsettia or Narcissus)*
  - Pearl Harbor Day
  - Hanukkah
  - Winter Solstice
  - Christmas
  - Kwanzaa

Source: Contributed by P. Catlin

Sample planting schedule for holidays

Holiday	Plant product	Propagation or preparation method	Timing (weeks prior to event)
Valentine's Day	Ivy heart topiary	Tip cuttings	4
St. Patrick's Day	Decorate pots of shamrocks	Plant corms or division	8

Holiday	Plant product	Propagation or preparation method	Timing (weeks prior to event)
Mother's Day	Blooming marigold pot	Seed	8
Father's Day	Succulent dish garden	Division and tip cuttings	4
Fourth of July	Red, white, and blue garden flower arrangements	*Statice*—seed *Celosia*—seed Dusty miller—buy transplants	*Statice*—16 *Celosia*—12 Dusty miller—6
Fall equinox	Fall leaf art	Press	2–3
Halloween	Spider planter	Offsets	2
Thanksgiving	Pumpkin planter (Swedish ivy or other hanging plants)	Tip cuttings	3
Winter holidays	Blooming amaryllis and paper white narcissus	Forcing bulbs	8–10

Bringing the outdoor garden indoors

Plant name	Cuttings	Seeds	Pressing and drying	Drying and teas	Potpourri, oils, and so on	Cook	Oils and vinegars
Begonia sp.	X						
Celosia sp.		X	X		X		
Coleus sp.	X						
Herbs	X	X		X	X	X	X
Everlastings— e.g., *Helichrysum, Limonium*		X	X		X		
Impatiens sp.	X						
Lemon balm—*Melissa officinalis*				X	X	X	X
Marigold— Tagetes sp.		X	X		X		
Mint— *Mentha* sp.	X			X	X	X	X
Nasturtium— *Tropaeolum majus*		X		X		X	
Geranium— Pelargonium sp.	X		X				
Rose—*Rosa* sp.		X			X		
Scented geranium	X		X	X	X	X	X
Inch plant— *Tradescantia* sp.	X						
Verbena sp.			X				
Vegetables		X				X	
Zinnia sp.		X	X				

Source: Contributed by P. Catlin

Print resources for horticultural therapy task and activity ideas

- *AHTA Newsmagazine.* www.ahta.org
- Bruce, Hank and Jill Folk. 2004. *Gardening Projects for Horticultural Therapy Programs.* Sorrento, FL: Petals & Pages Press.
- Cassidy, Patty. 2011. *The Illustrated Practical Guide to Gardening for Seniors.* Leicestershire, UK: Anness Publishing.
- Cassidy, Patty. 2013. *The Age Proof Garden: 101 Practical Ideas and Projects for Stress-Free, Low-Maintenance Senior Gardening.* Leicester, UK: Anness Publishing.
- Catlin, Pam. 2012. *The Growing Difference: Natural Success through Horticultural-Based Programming.* Create Space, doi: 9781477429662.
- Etherington, Natasha. 2012. *Gardening for Children with Autism Spectrum Disorders and Special Needs: Engaging with Nature to Combat Anxiety, Promote Sensory Integration and Build Social Skills.* Philadelphia, PA: Jessica Kingsley.
- Gabaldo, Maria M. et al. 2003. *Health through Horticulture: A Guide for Using the Outdoor Garden for Therapeutic Outcomes.* Glencoe, IL: Chicago Botanic Garden.
- Hewson, Mitchell L. 1994. *Horticulture as Therapy: A Practical Guide to Using Horticulture as a Therapeutic Tool.* Guelph, ON: Greenmor Printing.
- Hoetker Doherty, Janice. 2009. *A Calendar Year of Horticultural Therapy: How Tending Your Garden Can Tend Your Soul.* Boynton Beach, FL: Lilyflower Publishing.
- Jiler, James. 2006. *Doing Time in the Garden: Life Lessons through Prison Horticulture.* Oakland, CA: New Village Press.
- Larson, Jean and Mary Meyer. 2006. *Gardening Together: Sourcebook for Intergenerational Therapeutic Horticulture.* Binghamton, NY: Haworth Press.
- *Making Connections: A Newsletter from the Horticultural Therapy Institute.* www.htinstitute.org.
- Molen, Stephanie et al. 1999. *Growth through Nature: A Preschool Program for Children with Disabilities.* Sagaponack, NY: Sagapress.
- Moore, Bibby. 1989. *Growing with Gardening: A Twelve-Month Guide for Therapy, Recreation and Education.* Chapel Hill, NC: UNC Press Books.
- Rothert, Eugene and Kelly Nelson. 2011. *Health through Horticulture: Indoor Gardening Activity Plans.* Glencoe, IL: Chicago Botanic Garden.
- Wise, Joanna. 2015. *Digging for Victory: Horticultural Therapy with Veterans for Post-Traumatic Growth.* London, UK: Karnac Books.

Horticulture print resources for horticultural therapy task and activity ideas

The following are a few references on horticulture. These are useful for planning and growing within a variety of horticultural therapy programs, perhaps especially those programs that operate with a business model and/or organized sessions based on crops and growing schedules.

- Bartholomew, Mel. 2013. *All New Square Foot Gardening. 2nd edn.* Minneapolis, MN: Cool Springs Press.
- Coleman, Eliot. 1995. *The New Organic Grower: A Master's Manual of Techniques for the Home and Market Gardener. 2nd edn.* White River Junction, VT: Chelsea Green Publishing.
- Davidson, Harold, Roy Mecklenburg, and Curtis Peterson. 2000. *Nursery Management: Administration and Culture. 4th edn.* Upper Saddle River, NJ: Prentice Hall.
- Dole, John M. and Harold F. Wilkins. 2004. *Floriculture: Principles and Species. 2nd edn.* Upper Saddle River, NJ: Pearson/Prentice Hall.
- Ellis, Barbara. 2013. *Starting Seeds: How to Grow Healthy, Productive Vegetables, Herbs, and Flowers from Seed.* North Adams, MA: Storey Publishing.
- Fortier, Jean Martin. 2014. *The Market Gardener: A Successful Grower's Handbook for Small-Scale Organic Farming.* Gabriola Island, BC: New Society Publishers.
- Hartman, Hudson T. et al. 2010. *Hartmann & Kester's Plant Propagation: Principles and Practices. 8th edn.* Upper Saddle River, NJ: Prentice Hall.
- Jeavons, John. 2012. *How to Grow More Vegetables. 8th edn.* Berkeley, CA: Ten Speed Press.
- McMahon, Margaret et al. 2002. *Hartmann's Plant Science: Growth, Development, and Utilization of Cultivated Plants.* Upper Saddle River, NJ: Prentice Hall.
- Nelson, Paul V. 2011. *Greenhouse Operation and Management. 7th edn.* Upper Saddle River, NJ: Prentice Hall.
- Smith, Shane. 2000. *Greenhouse Gardener's Companion.* Golden, CO: Fulcrum Publishing.
- Still, Steven M. 1994. *Manual of Herbaceous Ornamental Plants. 4th edn.* Champaign, IL: Stipes Publishing.
- Stone, Curtis Allen. 2016. *The Urban Farmer: Growing Food for Profit on Leased and Borrowed Land.* Gabriola Island, BC: New Society Publishers.
- Whiting, David. 2012. *The Science of Gardening.* Dubuque, IA: Kendall Hunt Publishing.

Horticultural therapy session plan

Group Name: _____Session Date: _____

Program Model Type: _____

Goals: _____

_____

_____

Methods/Activity Summary: _____

_____

_____

Supplies: _____

_____

_____

Procedure: _____

_____

_____

_____

_____

_____

Documentation Method: _____

_____

_____

_____

Review and Follow-Up Notes: _____

_____

_____

_____

Source: Contributed by K. Kennedy for the Horticultural Therapy Institute

Activity description and planning tool: Terrariums

Time Allowed: 1 hour

Goals:

1. Maintain fine motor skills
2. Develop/maintain cognitive skills
3. Develop/maintain appropriate socialization skills

Objectives:

1. Participants will utilize fine motor skills for at least 10 min.
2. Participants will maintain alertness during 5–10 min of discussion time during session.
3. Participants will work cooperatively with others for the duration of the project.

Materials Needed:

- Clear plastic "suitable for terrarium" containers
- Pencils/pens for poking holes
- Potting mix and trays or bags
- Gloves
- Aquarium gravel
- Charcoal
- Suitable plants (ivy, *Fittonia*, or *Peperomia*)
- Water, warm and cool
- Scissors

Procedure:

1. Show an example of a terrarium and/or pictures to the group. Lead the group in a discussion of terrarium facts (see Points of Interest). Describe the project for the day and the plants that will be used.
2. Distribute the terrarium containers to the groups.
3. Cover the bottom of the terrarium with at least 1 inch of aquarium gravel.
4. Cover the gravel layer with a thin layer of charcoal.
5. Mix the potting mix with warm water in a resealable bag until evenly moist.
6. Cover the gravel and charcoal with a 2 inch layer of potting mix pressed in firmly.
7. Determine where plants are to go (two or three plants) and make a hole for each.

8. Place cuttings in holes and firm the potting mix around the stems.
9. Water gently around each plant.
10. To remove soil from the walls of containers, either pour a tiny amount of water down the walls or take a tissue and wipe them out.
11. Put on a lid and place in an area with indirect light.

Planning Notes:

Points of Interest:

- Terrariums were first used by early explorers as a way of keeping plants alive when bringing them back to their homelands from far across the oceans. Seawater could not be used and drinking water needed to be saved for the crew.
- Moisture from the soil and plants evaporates and rises to the surface where it condenses and returns to the soil as rain in a continuous cycle.
- As long as the seal is tight, there is no need to add water in the future.
- Charcoal keeps the terrarium water "sweet." Did you know that charcoal is often used in water filters to create better tasting water?
- What makes a good terrarium container? Anything clear that can be sealed, such as aquarium tanks, canning jars, soda bottles, giant pickle jars, and giant water bottles.
- Terrariums need to be kept out of direct sunshine, as plants can burn from the intensity of the sun through the clear glass or plastic.

Adaptations:

- To encourage more socialization and cooperation among peers you can select teams and have each team cooperatively create a terrarium.
- Use ivy only with those who are safe working with toxic plants.

Source: Contributed by P. Catlin

Session review checklist

External Factors: Including setting, location, and other variables impacting participant comfort and participation.

__ Yes   __ No   Session time conducive to maximum participation and
                 alertness?

__ Yes   __ No   Temperature comfortable?

__ Yes   __ No   Transportation to and from group?

__ Yes   __ No   Location convenient?

__ Yes   __ No   External distractions?

__ Yes   __ No   Lighting adequate/appropriate?

__ Yes   __ No   Seating arrangement conducive to group process?

__ Yes   __ No   Participants arrive prepared for session (appropriate dress,
                 sun protection, personal needs addressed, etc.)?

__ Yes   __ No   Access to restrooms, water, tissues, assistive devices and so
                 on is available?

Session Process: Variables that impact group function and intended outcomes.

__ Yes   __ No   Task appropriate to age, gender, and group ability?

__ Yes   __ No   Barriers to maximum participation levels addressed?

__ Yes   __ No   Ample amount of supplies?

__ Yes   __ No   Sufficient and appropriate tools?

__ Yes   __ No   Modifications and accommodations made efficiently and
                 appropriately?

__ Yes   __ No   Sufficient staff and/or volunteer support available?

__ Yes   __ No   Safety issues eliminated?

__ Yes   __ No   Adequate time for tasks?

__ Yes   __ No   Adequate time for discussion?

Self Review: Evaluation of the therapeutic use of one's self.

__ Yes   __ No   Directions presented at an appropriate level?

__ Yes   __ No   Directions presented in a logical order?

__ Yes   __ No   Tone of voice, choice of words, and interactions respectful?

__ Yes   __ No   Leadership style effective?

__ Yes   __ No   Able to focus on both the individual and the group?

__ Yes   __ No   Facilitated client's understanding of group purpose,
                 individual goals, and progress?

__ Yes   __ No   Pace of the group was comfortable for everyone?
__ Yes   __ No   Interaction between participants well managed?
__ Yes   __ No   Desired outcomes met?
__ Yes   __ No   Recognized appropriate client/therapist boundaries?
__ Yes   __ No   Demonstrated least restrictive training techniques
                 (demonstration, verbal cues, physical cues, and physical
                 assist) appropriately?
__ Yes   __ No   Promoted independence?
__ Yes   __ No   Created an atmosphere of trust, respect, and patience?
__ Yes   __ No   Created positive rapport with participants?

Source: Contributed by K. Kennedy

Appendix III: Forms for documentation of horticultural therapy treatment

The following are a few examples of forms used for various purposes and types of programs in horticultural therapy, including those used for assessment, referral and participation records, treatment planning, session and individual observation, and discharge.

Hanna University Geropsychiatric Center
Situational Assessment Observations

Client:

Session Activity: **Date:**

Observation	Rating	Comments
Attitude Toward Task		
Willing to engage		
Shows initiative and motivation		
Responsible with tools/plants		
Works until task completion		
Follows safety precautions		
Punctual/effective time management		
Understands task purpose		
Seeks help appropriately		
Emotional Control		
Patient/delays gratification		
Tolerates frustration appropriately		
Modulates own mood		
Shows impulse control		
Focuses on positives		
Manages anxiety		
Cognition/Task Performance Skills		
Understands/recalls instructions		
Attends to task/topic		Attention span: _____ minutes
Follows multistep directions		
Follow appropriate sequence		
Aware of own errors		
Can organize own task		
Able to problem solve		

Functions best with: (circle) Written instructions   Demonstration   Visual cues
            Constant cuing   Physical assist   Hand over hand
Adaptive tools/techniques: (describe)

Communication/Interpersonal Skills	Rating	Comments:
Socializes/tolerates peers		
Cooperates with others		
Accepts supervision/assistance		
Shares tools/equipment/space		
Initiates interaction		
Responds in interaction		
Shows flexibility/tolerates change		
Behaves appropriately		
Shares own experiences/feelings		
Shows self-confidence		

Physical Skills/Abilities	Rating	Comments
Understandable speech		
Adequate hearing		
Adequate vision		
Adequate sitting/standing balance		
Adequate gross motor skills		
Adequate fine motor/eye–hand skills		
Adequate muscle strength		
Adequate endurance/energy level		

Independent mobility with: (circle)

Ambulates independently   Crutches   Cane   Walker   Rollator   Wheelchair   Motorized wheelchair

Client Strengths:

Client Limitations:

Future Needs:

Rating Scale:

1 = Able/independent

2 = Able with structure/supervision

3 = Able with minimal physical/cognitive assistance

4 = Able with significant physical/cognitive assistance

5 = Unable

Source: Contributed by S. Sieradzki for Hanna University Geropsychiatric Center

Craig Hospital Horticultural Therapy Referral and Patient Participation

Traumatic brain injury: *Spinal cord injury:*

*LOCF:*_____   *Injury Level:*_____

*Comments*_____   *Comments*_____

Gardening Interests:

- Indoor (houseplant care/propagation)
- Outdoor (vegetables/flowers)
- Adaptive gardening tools
- Horticultural crafts
- Other:

Goals:

1. _____
2. _____
3. _____
4. _____

Discharge Date:_____ City/State: _____

Attending Physician:_____ Medical Record #: _____

Referred by:_____ Date of Referral:_____

Horticulture Group Participation:

Date: _____ Activity:_____

Reviewed Videos:

Date:

_____   Intro to Square Foot Gardening

_____   Square Foot Gardening Volumes 1, 2, and 3

_____   How to Grow Cool-Weather Vegetables

_____   Hydroponic Gardening

_____   Other

Horticultural Therapy Activities: In-House and Outings

Date: _____ Activity:_____

Written Resources

- *Raised-Bed Gardening*
- *Adaptive Garden Equipment*
- Tool Catalogue Vendor List
- Garden Plot Referral
- Review *The Enabling Garden*
- Garden Design Illustrations
- Other

Adaptive Equipment:

- Wide U-cuff (S M L)
- Forearm cuff (S M L)
- Submitted Request
- Other

Comments/Recommendations:

Discharge Note:

Source: Contributed by S. Hall for Craig Hospital

Mountain Valley Developmental Services Individual Service Support Plan (ISSP)

Person:
Start Date: **Program:** Vocational
Completion Expected: **Actual Completion Date:**

Short-Term Goal: A. will work up to 3 hours in the greenhouse per day.
  Objective 1: A. will produce 1 hour's worth of work.
  Objective 2: A. will produce 2 hours' worth of work.
  Objective 3: A. will produce 3 hours' worth of work.

  Criterion: 80% for 6 consecutive months, objective 1. Then A. will move to objective 2, and so on.

Prerequisite Skill: Desire to participate in more greenhouse activities, knowledge of tasks.

Current Performance: A. has difficulty staying on task. He will start working, then he will easily become distracted and fall off task.

Purpose of the Program: To help A. start a work-related task and finish the task in a timely manner.

Materials Needed: Supplies and equipment as needed to do tasks, staff encouragement.

Person(s) Responsible: Greenhouse staff, A.

Timelines for Review:
- Program Manager: Weekly and/or monthly
- Case Manager: Bi-annually
- BPRC/HRC: n/a

Steps (task analysis/behavioral interactions):
1. Staff will give A. the tools necessary to accomplish the work presented to him
2. Staff will help A. stay on task with verbal prompts
3. Staff will help A. build his capacity to work for longer periods of time

Reinforcement: Verbal praise, increased job skills, and paycheck

Correction Procedure: Verbal reminders.

Data Collection Procedure:
  + Task completed during day
  – Task not completed
  1 No opportunity

_____   _____
Signature/Date   Signature/Date

Source: Provided by B. Scrimsher for Mountain Valley Developmental Services

Sample Individual Treatment Plan and Methods for Horticultural Therapy

Person: Clientname   **Person Responsible:** Name of therapist
Start Date: **Program:** Therapeutic
Completion Expected: **Actual Completion Date:**

Long-Term Goal: Improve control over impulsive behaviors.

Short-Term Objective: Work cooperatively with peers and leader during horticultural therapy session
for four consecutive sessions.

Methods:
Sessions take place at gardens located at the residential treatment center. Sessions are held twice a week for
two-hours each time.

The horticultural therapist (HT) constructs sessions so that each one includes work in a small group, during
which time a specific task needs to be accomplished as a team. At the beginning of the session the therapist
could conduct a role-play of the day's activity. This role-play would demonstrate both cooperative and
uncooperative behaviors and some strategies for dealing with them, as well as introduce the task to be
accomplished during that session.

The HT monitors cooperative behavior throughout the session using a pre-agreed-upon three-strikes policy.
Initially, when the client demonstrates uncooperative group behavior (see note below for specific behaviors
warranting attention), the HT warns the client using a pre-agreed-upon hand signal or calling the client's name.
If undesired behavior continues, the HT takes the client aside and provides reminders about working cooperatively,
asks for a brief explanation of what is going on, and asks the client to come up with a positive way to approach
the group or peer. The HT may need to provide prompting or suggestions. If the behavior continues, client is
removed from the group for the rest of the session. (Editor's note: Removal from the group might encourage the
undesired behavior, so the HT should notice and discuss this with the client and modify the procedures if
needed.)

Note: The following behaviors are targeted for remediation, and would be cause for prompting.
- Physical altercation
- Verbal dispute (shouting, name calling, disruptive behavior)
- Lack of group participation (withdrawal)
- Uncompromising behavior

Documentation/Charting:

Date	Rating	Comments

Rating scale: 4= No prompting needed to accomplish objective
3 = One prompt
2 = Taken aside (two prompts)
1 = Removed from horticultural therapy session (three or more prompts)

Source: Contributed by Gail Doesken

Sample Group Single-Session Observation Mesh Chart Checklist

Deaconess Hospital of Cleveland Horticultural Therapy Program

Date: _____ Group: _____

Session Activity: _____

CLIENT NAMES:									
Followed and recalled multistep instructions									
Able to organize task in proper sequence									
Able to see and correct own errors									
Followed safety precautions									
Demonstrated emotional self-control									
Interacted positively with staff and peers									
Shared own past feelings/experiences									
Worked cooperatively with others									
Willing to engage for entire group session									
Showed flexibility/tolerated change									
Tolerated frustration appropriately									
Had adequate motor skills/coordination									
Had adequate muscle strength/endurance									
Worked independently with minimal cues									

Additional Notes: _____
_____
_____

Therapist Signature: _____ Date: _____

Source: Contributed by S. Sieradzki for Deaconess Hospital of Cleveland

BASICS Documentation

The following daily evaluation form was developed for the Horticultural Training Center at Father Flanagan Home for Boys (Boys Town) in Nebraska, The BASICS system of daily evaluation of youth employment focuses on positive advancement. Within a 2-week pay period, each employee had the opportunity to earn their "basics" on each working day. BASICS is an acronym for: **B**eing on time, **A**ttire, **S**taying on task, **I**nstruction following, **C**ondition of work, **S**ocial skills. Dependent on the percentage of excellence scores and competency points earned, employees were rewarded with raises.

Pay Period for the Weeks Of:

Employee	M	T	W	Th	F	M	T	W	Th	F	Total Hours
J.S.	4	4.5	4	3.5	2	4	4	4.5	4	4.5	38
	+B	+ B	+ B	+ B	+ B	B	+ B	+ B	+ B	+ B	
	+A +S	+ A	+ A	+ A	+ A	+ A	+ A	+ A	+ A	+ A	
	I	+ S	+ S	+ S	S	+ S	+ S	+ S	+ S	+ S	
	+C +S	+ I	+ I	+ I	+ I	I	I	+ I	+ I	+ I	
		+ C	+ C	+ C	+ C	+ C	+ C	+ C	+ C	+ C	
		+ S	+ S	+ S	+ S	+ S	+ S	+ S	+ S	+ S	

Being on time 90
Attire 100
Staying on task 90
Instruction following 70
Condition of work 100
Social skills 100
% of Excellence 550/600   (91%)

CONGRATULATIONS J.S.!!! You have successfully kept your percentage of excellence above 90% for six pay periods. Your next check will reflect a pay raise of $____ /hr.

Source: Contributed by T. Kent Titze, HTM

Sample Discharge Summary

Deaconess Hospital of Cleveland Horticultural Therapy Program

Client Name: Medical Record Number:
Diagnosis: Date of Referral:

Discharge Summary:
Discharge Date:

Client participated in _____ of _____ scheduled horticultural therapy sessions during this hospitalization.
Level and quality of participation were:

Horticultural Therapy Goal Status:

	Achieved	Minimally achieved	Partially achieved	Not achieved	Comments
Goal 1:					
Goal 2:					
Goal 3:					
Goal 4:					

Recommendations/Comments:

Therapist signature: _____   Date: _____

Source: Contributed by S. Sieradzki for Deaconess Hospital of Cleveland

Appendix IV: Horticultural therapy treatment strategies

The aim of this appendix is to be a helpful resource guide for horticultural therapists. It is designed to offer specific treatment techniques and strategies for a variety of programmatic settings and types.

Included are four program categories: mental health, physical health, vocational, and wellness. Each category is divided into three parts: a list of common diagnoses, treatment focus, and corresponding horticultural therapy strategies. The strategies (techniques) are tailored to meet the identified needs of the clients. The final category, wellness, is organized by dimensions of wellness (as described by the National Wellness Institute, 2015) rather than by diagnoses. The dimensions are occupational, physical, social, intellectual, spiritual, and emotional.

Note that some diagnoses are addressed in more than one program category, with corresponding strategies appropriate for that program type. It is worthwhile reviewing the strategies to address particular treatment issues for·individuals who present symptoms or issues without a formal diagnosis. For example, a child may exhibit behaviors similar to those with Attention Deficit Hyperactivity Disorder (ADHD), but a formal diagnosis may not have been made. The treatment areas, strategies, and techniques indicated may be apt for this child.

Source: Compiled by Heather G. Benson

Mental Health (*Contributed by S. Sieradzki and C. LaRocque*)

Diagnosis	Treatment focus	HT strategies
ADHD	Distractibility, irritability, argumentativeness	Assign detailed tasks to promote focus, integrate body and brain through whole-body activities (digging, walking, and lifting).
		Improve mood and self-esteem by offering choice of activities.
Oppositional-defiant disorder (ODD)	Difficulty taking responsibility for actions, sense that one should be able to control situations and other people's actions, and frustration when one cannot	Encourage responsibility and nurturing response through sustained plant care.
		Channel need for control into constructive activities such as planning the layout and choosing plants for an individual garden space.
		Facilitate opportunities for problem-solving with peers by also planning a group garden.
	Safety issues: Verbal or physical aggressiveness, refusal to follow directions	*Precautions:* Set and maintain limits fairly and calmly. Know and use safety protocol if clients become escalated or aggressive.
Anxiety disorders	Excessive worry, persistent negative and fearful view of the surrounding world, thinking distortions	Divert attention away from worrisome thoughts—plants are interesting and naturally motivating.

Physical restlessness and psychomotor agitation, physical discomfort caused by panic responses (rapid shallow breathing, muscle tension, and rapid heart rate)	Use relaxing background music.
	Allow physical outlets such as frequent brief breaks to stretch or walk through the greenhouse or garden.
	Encourage calming and comforting rhythmic actions: mixing soil, digging, and weeding.
	For extreme anxiety, use weighted materials, such as placing a heavy soil tray on lap.
Tremendous effort to avoid situations that might produce anxious sensations and worrisome thoughts	Blow dandelion seed heads to practice slow breathing techniques.
	Offer plant material choices to increase sense of control.
	Identify plant stress (drought, overwatering, constricted roots, lack of light).
	Anthropomorphize specific plants and state how they might be "feeling." Assist in modulating feelings in response to anxiety rather than trying to escape discomfort.

Mental Health (*Contributed by S. Sieradzki and C. LaRocque*)

Diagnosis	Treatment focus	HT strategies
Depressive disorders	Depressed or irritable mood, low energy, lack of interest in activities, low self-esteem, sleep disruption, hopelessness about future, overwhelming or numb emotional state, sense that one is not in control of one's life	Prepare plants to give as gifts.
Bipolar disorder		Use plant care as a metaphor for self-care: "What do both we and plants need to survive and thrive?"
Major depression		Use plant care to gain a sense of accomplishment.
		Offer choices of materials to foster control.
	Low motivation and energy level, lack of pleasure in usual activities, difficulty initiating tasks due to anhedonia	Allow the client to observe horticultural activity to start with.
		Offer interesting, colorful, and/or fragrant plants with which to work.
		Encourage engagement in one small step of the activity.
	Low self-esteem and limited self-efficacy	Provide positive encouragement by presenting plant activities that ensure success.
		Choose plants that can tolerate a wide range of handling and care, and that can be used for home care and/or gifts to others.
	Neglect of hygiene and grooming, impaired appetite and sleep	Offer plant care activities that reflect healthy care habits, such as dusting or washing leaves, deadheading, watering, fertilizing and transplanting.
	Safety issues: Suicidal ideation, impulsivity, self-injury	*Precautions:* Count and monitor tools and supplies before and after activity, use non-toxic plants, do not use sharp objects, use fiber pots instead of plastic or clay.

Schizophrenia and schizoaffective disorder	Delusions, thought disorder, tangentiality, loose associations, illogical thinking	Keep communication open and work to make a connection with the individual.
		Avoid correcting, criticizing, challenging, or confronting delusions. If delusional thinking causes disruptions, calmly redirect the person back to the topic or activity. If fearful, reassure the client that he or she is safe while here.
		Include a lot of green plants and be sure that materials are well organized in the activity space.
	Paranoia	Respect the client's personal space. Refrain from contact (handshake, hug) unless the client initiates it or gives permission.
		Position a garden seat or individual activity "station" that allows for a sense of enclosure, yet permits surveillance.
	Physical restlessness	Allow individuals to walk in and out of groups if possible.
		Lead the entire group in brief movement breaks that are based around gardening, such as moving plants into or out of the greenhouse.
	Cognitive deficits and disorganization	Simplify instructions: give one step at a time. Use multiple modes of instruction, both verbal and demonstrated. Seek feedback to check client's understanding. Limit the number of choices during activities. Be concrete in directions and discussion.

(Continued)

Mental Health (*Contributed by S. Sieradzki and C. LaRocque*)

Diagnosis	Treatment focus	HT strategies
	Extreme sensitivity to slights and criticism and vulnerability to stress	Use neutral, positive, and accepting speech, facial expressions, and body language.
		Use plants to illustrate responses or gentle directions.
	Hypersensitivity to sensory stimulation, inability to filter distractions	Reduce environmental sensory stimulation. Eliminate clutter.
		Provide a "green" space with a simple and clear design.
	Impulsivity and poor social skills, monopolizing conversation, interrupting/talking over others	Redirect to the horticultural activity at hand.
		Pass a plant or other garden object to each person to indicate that it is his or her turn to speak.
Substance use disorders	Irritability and restlessness due to the detoxification process	Assist client to develop new meaningful leisure skills and hobbies.
		Gardening provides opportunities for connecting with nature, spiritual practice, and caring for and healing the earth.
	Low tolerance for difficult emotions, impulsivity	Identify emotions and recognize their impermanent nature.
		Encourage counting to three before beginning the task.

	Poor problem-solving skills, tendency to quit activities, frustrates easily	Offer simple problem-solving activities with plants, such as where to place containers for watering.
		Gradually increase the duration of gardening sessions.
		Set short, reasonable goals. Give positive feedback for effort, regardless of accomplishment.
Trauma based disorders—posttraumatic stress disorder (PTSD) and borderline personality disorder	*Safety issues:* Impulsivity, impaired body awareness, changes in blood pressure and heart rate	*Precautions:* Monitor use of sharp tools or materials; prevent their use as weapons.
		Offer gross motor skills (digging, lifting) to improve body awareness.
	Tremendous effort to avoid situations that might produce panic attacks	Identify gardening tasks or spaces that are soothing for each individual to provide outlets to modulate feelings in response to panic or anxiety.
		Offer choices and use language such as "when you're ready..." or "if you'd like..."
	Physical restlessness and psychomotor agitation, physical discomfort caused by panic responses (rapid shallow breathing, muscle tension, and rapid heart rate)	Allow physical outlets such as frequent brief breaks to stretch or walk in the garden.
		Encourage calming and comforting rhythmic actions, such as raking or sweeping garden areas.
	Hypervigilance. Possible triggers: unexpected or uncomfortable touch, loud noises, or sudden movements	Avoid triggers whenever possible.
		Use garden fountains to reduce outside noise disturbance.
		Ask permission before touching client.

Mental Health (*Contributed by S. Sieradzki and C. LaRocque*)

Diagnosis	Treatment focus	HT strategies
		Avoid sudden movements and approaching client from behind.
		Have client identify garden tasks or spaces to use if he or she becomes accidentally triggered.
	Safety issues: Suicidal ideation and self-injurious behavior (such as cutting self or swallowing materials)	*Precautions:* Monitor all tools and supplies—take a count at the beginning and end of a session. Use nontoxic plants. Do not use sharp objects. Carefully watch individual's use of small objects, which could be swallowed.
Traumatic brain injury—caused by accidents, illness, or cerebral vascular accidents (stroke)	Cognitive deficits: Impaired short-term memory, concrete thought processing, confusion, impaired problem-solving	Simplify instructions: give one step at a time. Use multiple modes of instruction, both verbal and demonstrated. Seek feedback to check client's understanding.
	Impaired safety judgment	Limit the number of choices during activities. Be concrete in directions and discussion.
		Increase support, structure, and supervision to ensure safety.
		Partner an individual with cognitive deficits with an individual who has a higher cognitive level of function or with an able volunteer.
		Use nontoxic plants.

Physical Health (*Contributed by S. Hall and M. Wichrowski*)

Diagnosis	Treatment focus	Horticultural therapy strategies
Left cerebral vascular accident (stroke)	Right-sided weakness or paralysis	Utilize the affected side as tolerated to maximize return.
		Use the affected side for functional assistance if no voluntary movement is present.
	Vision may be affected	Orient individual to garden or greenhouse layout if vision affected.
		Follow Americans with Disabilities (ADA) guidelines to ensure greatest accessibility to garden spaces for those needing wheelchairs, walkers, or canes.
Right cerebral vascular accident (stroke)	Left-sided weakness or paralysis (hemiparesis or hemiplegia)	Utilize the affected side as tolerated to maximize return.
		Use the affected side for functional assistance if no voluntary movement is present.
	Left side neglect or field cut	Initially place materials at midline of sight. During subsequent sessions, shift materials increasingly to the left while prompting for scanning behavior.
	Impulsivity	Structure activity so that the horticultural therapist has control over access to tools and materials. Instruct patient to verbalize the step to himself or herself, or to count to three before engaging in the next step of the activity.
	Safety issues: Tool safety is an important consideration when impulsivity is present	*Precautions*: Supervise tool use. Use a dibble instead of a finger when pointing to where to cut.

Physical Health (*Contributed by S. Hall and M. Wichrowski*)

Diagnosis	Treatment focus	Horticultural therapy strategies
Neuromuscular conditions:	Muscle weakness, tremors, decreased endurance, loss of fine motor skills, range of motion	Grade activity to challenge but not frustrate the individual. For example, use larger seedling plugs instead of small seedlings for transplanting into containers.
Multiple sclerosis (MS), Parkinson's disease,		Build-up or pad handles on gardening tools.
Guillain-Barré syndrome,	Visual difficulties in later phases of MS	Weights such as a heavier tool choice can help reduce tremor.
Amyotrophic lateral sclerosis (ALS),		Provide orientation and consistent placement of gardening tools.
Huntington's disease	*Safety issues*: Heat exposure, dehydration, injury	*Precautions*: Avoid overheating in hot weather. Garden during the cooler morning hours. Provide and encourage participants to drink water. Use extra caution with sharp garden tools when tremors and sensory impairment are present.
Orthopedic conditions:	Regaining mobility and resuming previous life roles	Co-treatment with physical therapist working on standing tolerance and ambulation while engaged in garden tasks.
Knee replacement, hip replacement, serious fractures, amputations	Pain	Activities such as cutting and arranging flowers provide sensory and cognitive distraction that can compete with pain signals.

Cerebral palsy	Weakness, muscle rigidity and contractions, decreased range of motion, decreased fine motor skills	Encourage independence through use of raised beds, enabling use of garden tools, and choosing durable plants with which to work.
		Grade activity to meet the ability of the individual. For example, allow more time to complete the transplantation of annuals into the garden.
Spinal conditions:	Impaired grip	Offer physical assistance as necessary with difficult steps.
		Build-up or wrap handles on garden hand tools to reduce slipping.
Trauma (incomplete, complete)		Insert tool handles into "Velcro cuffs" with pockets.
		Use levers as handles on hose bibs or sink faucets for easier use.
	Decreased strength	Provide lightweight, short-handled tools.
		Offer long-lever scissors or add extensions to handles of pruners.
	Difficulty regulating body temperature	While outside in hot temperatures, wear protective clothing and a hat, recommend periodic rest breaks in the shade, mist skin surfaces (face, neck, and arms) with water from a spray bottle to decrease body temperature, and stay hydrated.
	Photosensitivity	Some medications can cause greater sensitivity to the sun (rash, burning). Wear sunscreen, a hat, and/or clothing to cover skin.
	Lack of sensation	Sensation to hot and cold may be impaired or absent. Skin may burn without feeling it. In the sun, cover dark-colored surfaces that bare skin touches, such as the edges of a raised bed.

(Continued)

Physical Health (*Contributed by S. Hall and M. Wichrowski*)

Diagnosis	Treatment focus	Horticultural therapy strategies
Traumatic brain injury (TBI)	Limited upper extremity reach, decreased trunk balance	Utilize raised planter beds and containers for easier access.
	Impaired sensory systems	Utilize various plants to elicit a response: visual, smell, texture, auditory, taste.
	Visual deficits (field cut, decreased acuity)	Customize written information to be concise, large print, simple font (Arial), and on light-colored paper.
		Establish an "anchor" for field cut deficits. For example, in a gardening setting, "scan to the left until you see the edge of the container." Wear sunglasses for light sensitivity.
	Hemiparesis	Introduce one-handed tools, including pruners with a "cut-and-hold" feature.
	Impaired standing balance	Provide a seat from which to work in the garden (portable lawn chair, garden bench, scooter, or kneeling bench). Raised planter beds can also act as a stationary support for standing work.
	Safety issues: Injury to self or others	*Precautions*: Evaluate safe use of equipment, taking extra caution with "sharps" and power tools. Provide adequate supervision and train volunteers how to safely supervise.

Vocational (*Contributed by S. Gallagher, J. Gabriel, and R. Haller*)

Diagnosis	Treatment focus	HT strategies
PTSD	Inattention, distractibility, poor impulse control	Design the garden work and growing spaces to create a safe, predictable environment; minimize distractions.
		Offer engaging activities that require increasing levels of sustained focus. For example, activities might progress from planting seeds and thinning seedlings to making a terrarium or caring for a garden bed.
	Impaired reality testing	Use stimulating sensory activities to bring focus to the present reality. For example, harvest mint and make tea, pot up scented geraniums, or create flower arrangements.
	Depression, anger, anxiety, irritability, mood swings	Offer gardening tasks that develop the capacity to self-soothe, such as watering, deadheading, or weeding.
		Draw parallels from garden care to self-care.
	Medication effects (e.g., drowsiness)	Offer breaks as appropriate.
		Monitor use of sharp tools.

(Continued)

Vocational (*Contributed by S. Gallagher, J. Gabriel, and R. Haller*)

Diagnosis	Treatment focus	HT strategies
Developmentally disabled/autism spectrum disorders (ASD)	Difficulty sequencing and following multistep instructions	Demonstrate one step at a time to develop mastery.
		Offer clear and concise communication.
		Allow extra time for processing.
		Provide visual cue cards for task steps.
	Easily frustrated; tantrums and meltdowns	Redirect to the garden task.
		Build in periodic rewards to develop confidence.
		Provide appropriate praise frequently.
	Safety issues: Low frustration tolerance; may become aggressive with little or no warning	*Precautions*: Closely supervise use of power tools, such as lawn mower or trimmer, and garden hand tools (particularly if sharp).
	Ineffective social skills	Model and encourage healthy social skills such as working in a team, respecting space, sharing tools, and polite communication.
	Hypersensitivity to changes in the environment	Establish a garden routine and, when possible, involve the same staff.

	Minimize changes in tasks and work, using task analysis to be consistent in garden task sequence and methods.
	Provide ample advance notice of upcoming changes.
Over- or under-sensitivity to stimuli (textures, light, smells, sounds)	Use effective garden design to create a safe, structured, comfortable environment.
	Minimize environmental distractions by partitioning space with trellises, raised beds, walls, or plant materials.
Awkward gait, use of mobility devices	Provide spacious work areas with unobstructed pathways.
	Gently redirect attention to the task at hand.
Repetitive behaviors (e.g., rocking, hand flapping, clapping)	Engage in repetitive yet purposeful horticulture tasks such as transplanting or weeding.
Difficulty communicating and understanding verbal instructions	Offer one-on-one interactions.
	Provide simple, clear instructions; avoid using condescending tone or words.
	Be patient and avoid finishing a person's tasks and sentences.
Stroke, spinal cord injury, MS, Parkinson's disease, amputations	Sensitively inject humor into activities.
Depression, anger, frustration	Offer tasks that provide built-in rewards, such as potting up a plant. Create a garden meal or item to share, sell, or donate.

(Continued)

Vocational (*Contributed by S. Gallagher, J. Gabriel, and R. Haller*)

Diagnosis	Treatment focus	HT strategies
	Social withdrawal	Set up group activities that foster social interaction and peer-to-peer coaching.
	Loss of gross and fine motor skills	Conduct plant sales "staffed" by clients.
	Lack of stamina; easily fatigued	Model and demonstrate tool use, including adapted gardening tools to compensate for loss of function.
		Monitor and modify activity duration as needed.
	Use of wheelchairs and other mobility devices	Make sure garden is accessible, with unobstructed work areas and pathways.
	Medication-induced photosensitivity	Provide ample shade in garden work and break areas.
Traumatic brain injury	Memory loss; difficulty concentrating, sequencing, planning	Provide simple, clear instructions.
		Use checklists and kitchen timers to jog memory.
		Provide a well-organized, well-labeled workspace with minimal distractions. For example, planting containers should be stored in an organized fashion with sizes labeled and easy to access.
		Repeat tasks and activities during the learning process.

Depression, anxiety	Emphasize a "real" work environment with plant care as a core part of the daily routine.
Inattention to detail; impulsivity	Engage in activities that require problem-solving such as propagating seasonal plants, building a trellis, or planting a garden bed.
Safety issue: Impulsive behavior may cause injury	*Precaution:* Monitor use of sharp tools.
Easily fatigued; headaches, seizures	Be mindful of session length and offer frequent breaks.
Inattention, distractibility	Use session opening routines that highlight sensory awareness. Perhaps ask, "what do you see, hear, smell?"
Adolescent youth within the juvenile justice system	Offer activities that require sustained attention such as seeding, flower arranging, or creating garden labels or signs.
PTSD, depression, anxiety, anger, aggression, impulsivity	Engage in relaxing activities such as watering, deadheading, or potting up herbs.
	Engage in highly physical activities such as digging beds for planting, spreading mulch, or turning a compost pile.
Safety issues: Anger, impulsivity, and aggressive behavior may cause injury	*Precautions:* Train in safe use of sharp tools and monitor use closely.

(Continued)

Vocational (*Contributed by S. Gallagher, J. Gabriel, and R. Haller*)

Diagnosis	Treatment focus	HT strategies
	Low self-esteem, low self-efficacy	Engage in projects that involve research, creativity, or problem-solving such as designing and building a compost bin or creating a production schedule for growing the ingredients to make salsa for future sales.
		Have youth who have mastered tasks coach others new to the activities.
	Suspiciousness, distrust	Nurture trust by working with youth one-on-one or in small groups.
		Be straightforward, genuine, and nonjudgmental in communications.
	Difficulty focusing and following multistep instructions because of substance abuse	Provide simple, clear instructions without being condescending.
Incarcerated adults	Lack of purposeful activity	Provide activities that involve community service such as growing food for food banks or landscaping to beautify the prison/jail environment.
	Anger, depression, anxiety	Engage in productive and relaxing activities such as seeding, watering, weeding, deadheading, or flower arranging.
	Low self-esteem, low self-efficacy	Offer activities that develop a sense of mastery and self-confidence. Care independently for a garden bed, install an irrigation system, or mentor others in mastered tasks.

Safety issue: Injury to self or others	*Precautions:* Monitor use of sharp tools. Develop checklists for keeping track of tools and materials.
Distrust of interpersonal relationships	Be direct and honest in communications.
Short attention span and easily distracted	Divide tasks into small, achievable steps.
	Organize the garden or greenhouse space into distinct task areas.
	Create workstations away from excessive noise and visual distractions. Be intentional about location and/or design. Design may incorporate trellises, potting benches with backs, and so on.
Adults with Down Syndrome	
Difficulty in sequencing steps in a task and following complex instructions	After detailed training on a task, provide a pictorial chart for the client to check off each step as it is completed. For example, fill pot to rim, dibble hole in center, insert plant into hole at proper depth, firm potting mix around plant with fingers, water.
	Give verbal or other cues to prompt memory.
	Use simple concepts.
	Offer routine and repetitive tasks for mastery such as transplanting seedlings, washing pots, deadheading flowers, and putting tools away.

(Continued)

Vocational (*Contributed by S. Gallagher, J. Gabriel, and R. Haller*)

Diagnosis	Treatment focus	HT strategies
	Limited physical endurance Impaired fine and gross motor skills	Engage in physical activities with a purpose such as moving plants in and out of the greenhouse.
	Communication skills may be impaired	Offer visual demonstration of the task while speaking.
		Watch for visual clues to understand vocalizations of client.
	Low self-esteem Inappropriate social interactions	After allowing ample time for task mastery, encourage individuals to mentor each other.
		Redirect client to the garden task at hand. Give gentle, firm, and consistent responses to behaviors.
Adults with cerebral palsy	Limited gross and/or fine motor skills	Define a workplace that is accessible, comfortable, and safe.
		Offer adaptive tools.
	Communication skills may be limited	Model patience and creativity while using a variety of communication methods. Share that plants communicate subtly; we must be patient to "see" what they are communicating.
	Visual impairment	Keep pathways clear and direct.
		All tools and supplies should remain within sight and reach.

Wellness *(Contributed by K. Kennedy and S. Taft)*

Wellness dimension	Treatment focus	Horticultural therapy strategies
Occupational A sense of satisfaction and fulfillment resulting from meaningful activity; includes employment and volunteer service as well as other meaningful activities	• Functional, transferable skills • Finding personal meaning and purpose • Purposeful tasks • Meaningful and relevant tasks	Design sessions that produce a product or an activity that contributes to or improves the client's living environment or community, such as • Flower arrangements in public areas or dining rooms • Caring for gardens or indoor plants in community areas • Adopting and caring for a garden in a community park • Creating items from the garden to sell or donate • Hosting a garden or tea party in which the clients invite family, friends, staff, or children from on-site day care Do real, purposeful tasks in the garden that need to be done for the health of the garden.
Physical A personal level of physical fitness as well as healthy lifestyle behaviors	• Managing food, alcohol, and tobacco intake • Building strength, endurance, and flexibility • Taking responsibility for personal wellness	Utilize the garden to promote interest in healthy food choices: • Prepare and taste simple dishes made from garden harvest such as salsas, vegetable side or main dishes, and beverages enhanced with herbal flavors. Demonstrate and use adaptive tools and garden spaces to maximize independent participation. Gradually increase the size of or the amount of water in the watering can to improve strength. Work to increase gardening time without breaks to improve endurance (or standing time to improve standing or dynamic balance).

(Continued)

Wellness (*Contributed by K. Kennedy and S. Taft*)

Wellness dimension	Treatment focus	Horticultural therapy strategies
Social Contributing to one's community through harmonious personal relationships and friendships	• Interpersonal communication skills • Getting along with others • Problem-solving for the greater good • Isolation	Set up session dynamics to include opportunities for cooperation, eye contact, and the sharing of tools and materials. Make problem-solving and planning part of the session. Include discussion in the task or activity and focus on eye contact, speaking clearly, and other interpersonal communication skills. Promote partnering opportunities within the session. Incorporate familiar plants and have clients share personal experiences and preferences. Make tussie-mussies (nosegays) using the Victorian language of flowers to express the qualities that clients see in each other or to communicate a message to a caregiver or loved one. Host a tea party to practice appropriate social behavior. Create a fairy garden with everyone contributing to the overall design and finished product.

Intellectual Using creativity to stimulate the mind as well as building knowledge and skills	• Opportunity for creative outlet • Problem-solving • Acquiring new skills and knowledge	Incorporate creativity and choice-making by offering a manageable number of selections such as ribbon or flower colors, dish garden plants, type of label, and so on. Read directions on seed packets or care information on fact sheets to determine planting directions. Present the group with a task and determine the steps needed to accomplish it. Plan a themed garden. Select some herbs to study, and discover their historical uses and lore.
Spiritual The search or exploration of meaning and purpose in life and the understanding of something greater than ourselves	• Harmony between self and behavioral choices and personal beliefs • Harmony between emotions and personal beliefs • Tolerance of the beliefs of others	Explore personal goal-setting techniques and write individual life-enrichment goals. Explore the plants in an ecosystem such as a wooded area, field, or water's edge to explore the balance of harmony in nature and within oneself: • Look for interconnectedness in relationships among species. • Look for symbiotic relationships in nature. Plant a Three Sisters Garden. Plan and plant a meditation garden. Create a finger labyrinth using dried plant materials to outline the path.

(Continued)

Wellness (*Contributed by K. Kennedy and S. Taft*)

Wellness dimension	Treatment focus	Horticultural therapy strategies
Emotional Being aware of, accepting, and managing one's feelings; coping with stress and developing a sense of autonomy and independence	• Managing one's emotions • Expressing emotions in positive and acceptable ways • Stress management techniques • Managing interpersonal relationships	Increase awareness of the process of creating personal growth and change: • Include guided journaling as part of a session or as session follow-up. • Teach clients how to identify steps toward personal goals using accountability techniques. Improve awareness of stress management techniques: • Rate stress level before and after weeding in the garden (or other repetitive task). • Create personal tea or potpourri blends and discuss the impact of fragrances on feelings of well-being. Provide problem-solving and group interaction opportunities by assigning small groups to accomplish garden tasks. Use garden metaphors as catalysts to discuss relationship or emotional issues, such as • Using the rose to illustrate positive support from family and friends and the thorns to illustrate unsupportive words and behaviors. • Supportive environment and relationships demonstrated by the Three Sisters Garden. Pruning to illustrate the opportunities for new growth when overgrowth is removed or cut back.

Spiritual		

The search or exploration of meaning and purpose in life; the understanding of something greater than ourselves	• Harmony between self and behavioral choices and personal beliefs
	• Harmony between emotions and personal beliefs
	• Tolerance of the beliefs of others
		Explore the plants in an ecosystem such as a wooded area, field, or water's edge to explore the balance of harmony in nature and within oneself:
		  • Look for interconnectedness in relationships among species.
		  • Look for symbiotic relationships in nature.
		Plant a Three Sisters Garden.
		Plan and plant a meditation garden.
		Create a finger labyrinth using dried plant materials to outline the path.

Emotional		

Being aware of, accepting, and managing one's feelings; coping with stress and developing a sense of autonomy and independence	• Managing one's emotions
	• Expressing emotions in positive and acceptable ways
	• Using stress management techniques
	• Managing interpersonal relationships
		Increase awareness of the process of creating personal growth and change:
		  • Include guided journaling as part of a session or as session follow-up.
		  • Teach clients how to identify steps toward personal goals using accountability techniques.
		Improve awareness of stress management techniques:
		  • Rate stress level before and after weeding in the garden (or other repetitive task).
		  • Create personal tea or potpourri blends and discuss the impact of fragrances on feelings of well-being.

(Continued)

Wellness (*Contributed by K. Kennedy and S. Taft*)

Wellness dimension	Treatment focus	Horticultural therapy strategies
		Provide problem-solving and group interaction opportunities by assigning small groups to accomplish garden tasks.
		Use garden metaphors as catalysts to discuss relationship or emotional issues, such as
		• Using the rose to illustrate positive support from family and friends and the thorns to illustrate unsupportive words and behaviors.
		• Supportive environment and relationships demonstrated by the Three Sisters Garden.
		• Pruning to illustrate the opportunities for new growth when overgrowth is removed or cut back.

Note: Unlike the other parts of this appendix, which are organized by diagnoses, this section addresses the six dimensions of wellness as defined by the National Wellness Institute. These dimensions are applicable to all persons, regardless of the reason that prompted horticultural therapy treatment. For a definition and more information on wellness from this organization, see *The Six Dimensions of Wellness* (retrieved June 30, 2015). http://www.nationalwellness.org/?page=Six_Dimensions&hhSearchTerms=%22definition+and+wellness%22.

Index

Note: e, f, and t with folios indicates exhibits, figures, and tables respectively.

Printed in the United States
by Baker & Taylor Publisher Services

Printed in the United States
by Baker & Taylor Publisher Services